The Seasons

The Seasons

Rhythms of life: cycles of change

Anthony Smith

Weidenfeld and Nicolson
5 Winsley Street London W1

Designed by Shashi Rawal for
George Weidenfeld and Nicolson Limited

SBN 297 00201 5

Printed in Great Britain by Cox & Wyman Limited
London, Fakenham and Reading

Jacket illustrations:
front a solar flare
back above flowers which grow only above 11,000 ft;
back below desert landscape, Montana, USA

Contents

Acknowledgments

The author and publishers wish to thank all those listed below who have kindly contributed photographs for this book.

Aerofilms, 178
Associated Press, 228 bottom
Australian News and Information Service, 63 (4), 199 top
British Museum of Natural History, 62 (3)
British Overseas Airways Corporation, 269 top
Douglas Botting and John Bayliss, 28 (2), 65 (2), 106 (2), 107 top right
Camera Press, 142 bottom and endpaper
James Carr, 29 top, 232 top right, 107
Pamela Chandler, jacket back flap
John Collins, 228 middle
R. Cooke, 36 top left, 66 bottom, 68, 194 bottom
Bruce Coleman Ltd, 31 bottom (2), 34 bottom left, 101, 103 bottom, 104 bottom, 108 (3), 143 top and middle, 184 bottom, 194 top right, 195 (3), 196 top and middle right, and bottom 225 (2), 226 bottom, 222 (2), 228 (top), 229 (3), 230 top and bottom right, 232 top left and bottom, 265, 266 (4), 267 (2)
J. E. Downward, 271 (4)
Mary Evans Picture Library, 138 bottom
D. Harris, 66 top, 272 top right
Victor Kennet, 272 top left
Keystone Press, 26–7, 64 top left, 104 top
Frank W. Lane, 36 top right, 133, 136, 138 top, 139 (2), 140 bottom, 172, 173 (2), 174 bottom (2), 175 (2), 176–7 (3), 181, 183, 196 top left, 226 top, 268–9 bottom (5)
John Markham, 31 top, 30
The Meteorological Office, 179
Mondadori Press, 29 middle, 32–3, 61 (3), 141 bottom, 193, 268 top left, 272 bottom
Dr N. Mrosovsky, 230 bottom left, 231 top
Novosti Press, 64 bottom (2), 105 bottom
D. E. Pedgley, 67 top, 143 bottom
Picturepoint, 35, 67 bottom right, 134, 135, 142 top, 180, 270 (2)
Popperfoto, 34 top, 64 top right, 105 top right, 107 top left, 140 top, 171 bottom, 174 top, 194 top left, 268 top right

P. A. Smithson, 137
Sunday Times Magazine, 102, 28 bottom
Telegraph Colour Library, 141 top, 144 (2)
United States Information Service, 36 bottom
The Zoological Society, 29 bottom

List of Diagrams

The Seasons

Introduction

Lingering with all of us from our schooldays is a statement of exceptional bluntness *The planet Earth rotates on its axis once every twenty-four hours.* It is one of those phrases (and plenty more emanate from the class-room) which can obscure an issue without effort. Why should the Earth behave in this fashion? Do footballs rotate on their axis, or is it axes? And what else ought an Earth to rotate on? The miasma accompanying the phrase soon buries it beneath misunderstanding.

Nevertheless, the remark is crucial. By spinning like a top, and because only one bright sun is in the vicinity, the Earth has a relentless succession of nights and days. By spinning like a top at an angle to the sun, and by maintaining that angle while travelling around the sun, the Earth also has a steady succession of seasons. It is possible to imagine a different planet, which always keeps one half facing its sun, and such a planet would never have nights alternating with days. It is also possible to imagine yet another planet which revolves so that its equator is always nearest to the sun, in which case there would be no seasons as the Earth knows them. Instead, our own planet Earth revolves both fairly rapidly (to cause its nights and days), and slightly obliquely to the plan of its rotation around the Sun (to cause its seasons). The planet Earth does indeed rotate once every twenty-four hours. And it rotates upon its axis.

The axis of a top, spinning perfectly, runs from the point at which it touches the ground to the tip of its handle perpendicularly above. Therefore its axis is perpendicular to, or at 90 degrees to, the floor. Unless it tips over, or starts to run down and wobble, the axis will remain at 90 degrees to the plane of the floor. The spinning Earth is of course standing on no floor, but it travels round the sun in the same plane, and this plane has some of the effects of a floor. However, unlike the vertical top, the Earth is perpetually canted over at an angle of approximately 66½ degrees. This angle does not mean that the north pole, for example, is forever nearer the sun than the south pole. Instead its constancy means quite the reverse; for half the year the north pole is nearer the sun than is the south pole, and for the other half the south pole has its turn. Whichever hemisphere is canted over to face the sun is receiving more radiation, and is therefore having its summer.

Contrary to expectation, at least by those of us who live in the northern

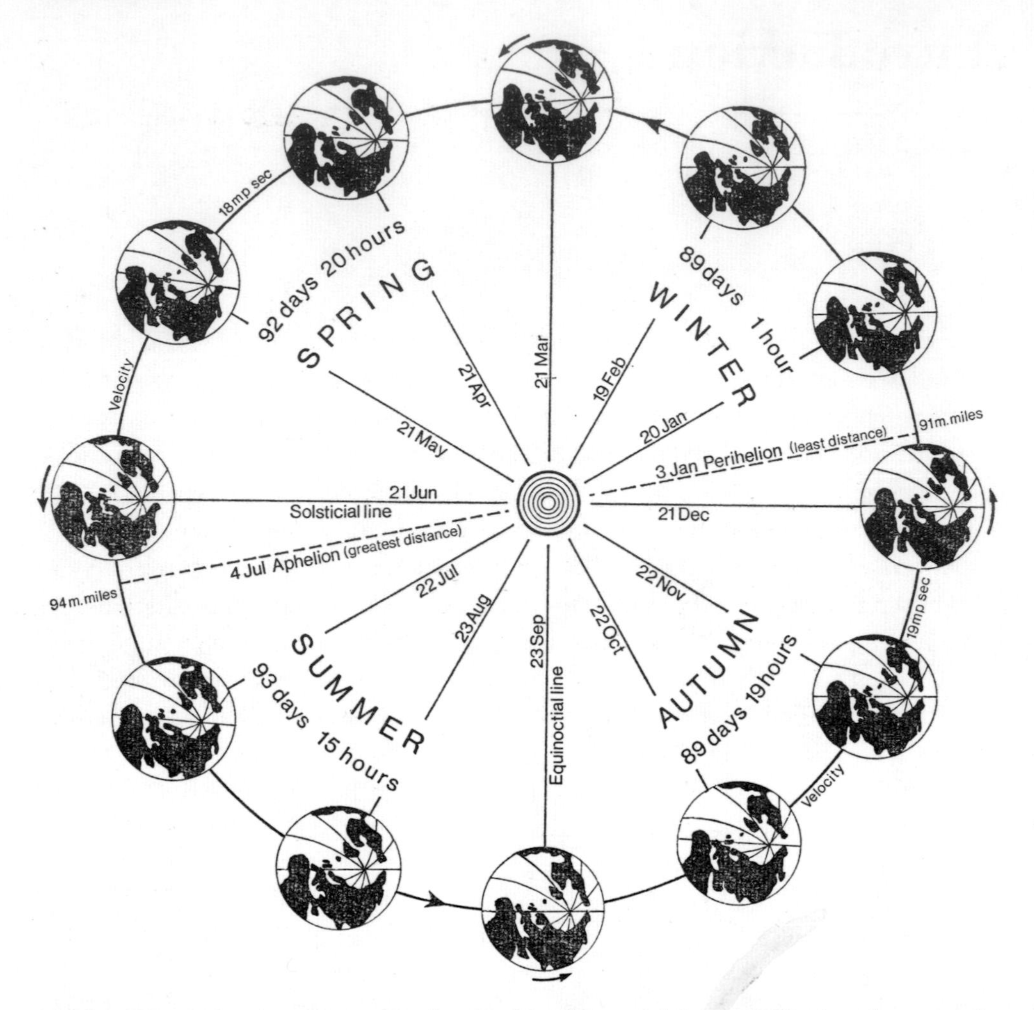

Northern Hemisphere seasons are plotted on Earth's orbit around the sun, falling into four not quite equal parts. Spring and Summer are each about 5 days longer than Autumn and Winter owing to the fact that the earth's path around the sun is slightly elliptical; the swing of roughly three miles away from the normal path occurs around 4 July, giving Spring and Summer an extra four days each.

hemisphere, the Earth is actually nearer the sun – by about three million miles – in the northern winter than in the northern summer. The actual date of perihelion, or nearest approach, is 2 or 3 January: the date of aphelion, or greatest distance, is 1 or 2 July. When it is nearer during midwinter the sun appears to be larger, but only by 1 minute 8 seconds of arc. Such a modest increase can only be detected by instruments, but in fact it causes a 7% increase in the solar radiation reaching this planet, a 7% increase in light and heat. Therefore, whatever northerners may think, the northern winter is less cold than the southern one, and less cold than it would be without this extra advantage. Similarly, the northern summer, due to that greater distance from the sun and the decrease in radiation, is less warm than it would be if the Earth's orbit occupied the same area but was circular.

If the Earth travelled round the sun in a perfect circle there would be no nearer or greater distance. Instead it travels in a slightly eccentric orbit. (Presumably no orbiting object anywhere in the heavens actually orbits in a circle rather than some form of eccentric ellipse.) Also, like all orbiting objects, it travels faster when nearer the body around which it is rotating. In January the planet Earth is travelling around the sun at about nineteen miles a second; in July the speed is nearer eighteen miles a second. These differences in speed help to explain why the four seasons are of different lengths, with the order in the northern hemisphere being summer, spring, autumn and winter from the longest to the shortest. This discrepancy is intriguing and needs elaboration, but first a word about the equinoxes and the solstices, the crucial markers of our seasonal year. They are its turning points.

The Earth is at one of its solstices when one of the poles is pointing as nearly in the direction of the sun as it will ever do. The famous axis on which the Earth spins never points directly towards the sun; instead, and at the northern hemisphere's summer solstice, the northern part of the axis is merely pointing more towards the sun than at any other time in the year. The date of this event is 21 June. It follows that on the same day the southern pole is facing more away from the sun than at any other time. The contrary event, when south pole and north pole have changed, and when the south pole's axis is pointing more nearly towards the sun than at any other time, occurs on 21 December. These two dates mark the start of summer and winter. Of course there have been hot summer days in the northern hemisphere before 21 June and wintry days before 21 December, but the solstices are astronomical markers rather than climatic datings. This complexity will form an integral part of this book because the seasons are caused by the astronomical situation but also describe the climatic consequences of this situation: the same word has to cover both events. Astronomers say that the northern winter begins on 21 December: all northerners would disagree, but would be incapable of agreeing on any particular date.

The equinoxes are the two other astronomical markers. On these occasions, 23 September and 21 March, the Earth's axis does not point towards the sun but is at right angles to it. Also, on these two days alone, both poles receive equal illumination from the sun, both day and night are of equal length (hence equinox) and, unlike all other days in the year, the sun is precisely overhead at midday on the equator.

Therefore, astronomically at least, summer lasts – in the northern hemisphere – from the summer solstice to the autumnal equinox (93·7 days); autumn lasts from its equinox to the winter solstice (89·6 days); winter lasts from its solstice to the vernal equinox (89 days); and spring lasts from its equinox to the summer solstice (92·9 days). These four seasons add up to an awkward fraction more than 365 days. A further difficulty is that the four seasons are themselves not static: the number of days quoted for each season is an average figure, not an implacably unchanging one.

The complex rhythm of our planet Earth, rotating upon its axis in its orbit around the sun, encircled all the while by its neighbour the moon, provides the basis for the ceaseless rhythm of life itself.

1 The Calendar

It is important to explain at the very start of this book how mankind has tried to date the year, both at its beginning and throughout its awkward passage of 365 and a bit days. Large numbers of people celebrate New Year's Eve, or New Year's Day; and, in doing so, feel that some great cosmic event is occurring. They are even prepared to argue that something astronomical, much like an eclipse, is happening at that important hour of midnight. It can actually be disquieting when television or radio sends a live transmission from another country in the throes of New Year junketings at some false hour like 6 p.m. our time. Revellers here and there are tinged with a sensation equivalent one feels to the emotion that must have touched the Stonehenge men when they watched the midsummer dawn first strike the crucial stone. Unfortunately, whereas midsummer is a most positive event, the hour of midnight on 31 December is nothing at all. It is no more cosmic than the Queen of England's official birthday or the start of the flat-racing season. It is entirely man-made.

Moreover, different men have different ideas about the start of the New Year. Most Europeans think it is 1 January. Jews have their own day somewhere between 5 September and 7 October. Moslems have a year lasting only 354 days and 8 hours, and therefore a yet more peripatetic New Year's Day. The ancient Greeks thought all new years began in the autumn. Others believed, and some still do, that the March equinox is a more reasonable beginning, and for centuries that time of the year was thought to be most auspicious. It was Julius Caesar who hit upon 1 January, that date being the first day of the first month after the winter solstice. Until then the Roman year had begun on 1 March, being the first day of the month containing the spring equinox (and the Celts and the Druids felt likewise) but in the revision of the calendar Julius Caesar brushed the old scheme aside. The Romans then so entrenched this notion of 1 January that much of the world today is stuck with it.

The Roman revision was thought necessary because of the inheritance of ideas that each year should be made up of a definite number of moon cycles, or months. The time taken by the moon to go round the Earth is, of course, entirely independent of the time taken by the Earth to go round the sun; but many prehistoric and early men could not resist the temptation to bracket the two together. All annual cycles were therefore composed of lunar cycles, and the price paid for this presumption was a need for eternal modifications to make things fit.

Originally, the Romans, for example, had ten months to the year: March, April, May, June, fifth month, sixth month, seventh month, eighth month, ninth month and tenth month. Then, as these ten moon cycles proved inadequate, two more were added – January, at the beginning, and February at the end. In 452 BC February was promoted to follow January, and this addition of two months at the beginning meant that all the numbered months were wrongly labelled. Even today, September, our ninth month, and all others to December, our twelfth month, are two months later than their names indicate. The twelve Roman months at that time were originally of twenty-nine and thirty days alternatively, trying to fit in with the lunar cycle. The Romans also threw in one odd day making a basic total of 355 days per year. This prevailing shortness was made good every other year by the insertion of another month – Mercedonius – between 23 and 24 February, and shortly before their New Year's Day. This thirteenth month was alternatively of twenty-two and twenty-three days, and had the effect of making an average year last too long – for 366¼ days. Every twenty-four years, due to this error, things were readjusted to try to make the calendar correct. It was this muddled situation which Julius Caesar inherited, and then determined to change.

First and foremost he discounted the idea that a year should be composed of lunar cycles: the lunar month was therefore replaced by the calendar months. (Unfortunately the word month has been retained, thus perpetuating the fallacy that a month is still something to do with the moon.) Julius Caesar also decided that all the alternate months of January, March, May, July, September and November should have thirty-one days (no lunar cycle there) whereas the months in between should have thirty. The month of February, always the afterthought month, was to be an exception, having twenty-nine days normally but thirty days every fourth year. Thus there were 365 days in an ordinary year, 366 in each fourth year, and an average of 365¼ days per year. The first so-called Julian year was 46 BC; the year before was called the last year of confusion because it contained the twelve traditional months, plus the extra month Mercedonius, plus two extraordinary months, totalling 445 days. The next year shrank to an orderly, and astronomically valid, amount.

Having achieved a satisfactory 365¼ days in a year the emperor relaxed. As reward, his name replaced Quintilis, the name of the original fifth month, and July is now our inheritance. Unfortunately the Emperor Augustus, not to be outdone in honours by his predecessor, gave his name to the sixth month. This sixth month, to be called August, was one of Julius's short months, and therefore inferior to Julius's thirty-one-day July. Augustus raised August's total to thirty-one, but this meant that July, August and September were three long months in succession. This was thought to be unsatisfactory, and he therefore changed the alternating system, in so far as this was now feasible, by ordaining that September and November should be reduced to thirty days and that October and December should be raised to thirty-one. February – forgotten month! – lost another day to make things come out right again at 365, with 366 every fourth year. Therefore the villain in the story is Augustus who so promptly destroyed the Julian precision and replaced it with a totally disorderly arrangement. Only the most brilliant among us can remember the system without recourse to inane and arhythmic jingles. We mouth 'Thirty days hath November, April, June and September. February hath twenty-eight alone, and all the rest have thirty-one' – when we should also be pouring scorn on Caesar Augustus.

The calendar months are thus entirely man-made, but even Julius, or rather his Greek astronomer Sosigenes, had made an imperfect system. Each Julian year was, on average, wrong by slightly over eleven minutes. The accumulated error was corrected in AD 325 by Constantine the Great when three days were omitted from the calendar. By AD 1582, Pope Gregory XIII decided that ten days should be lost to make things good again, and decreed that 1582 should be a mere 355 days long. Such papal insolence, and such a drastic move, was resisted in many parts. Britain resisted it for nearly 200 years, thereby compounding Julian's error still further, but in the year 1752 there were no British days between 2 September and 14 September. The lost 264 hours caused riots by those who felt cheated of eleven days of life. (Scotland was unaffected by the disturbance as the change there had been quietly made in 1600.) The lost days are responsible for some modern peculiarities. Each financial year used to end on 25 March; the new style made this 5 April. Common fields of the manor used to be open for grazing on Lammas Day, 1 August: new style made this 12 August, and a legacy of this postponement is that grouse-shooting in Britain begins on that date.

Gregory not only put things right, but made the leap-year arrangement more accurate than before. The Earth does not take 365¼ days to go round the sun, but 365 days, 5 hours, 48 minutes and 46 seconds. Therefore a fault in our calendar is almost inevitable. Pope Gregory did his best. He removed three days in every 400 years by decreeing that only the centuries which were multiples of 400 should be leap years. Thus 1900 was not a leap year: 2000 will be. To make everything yet more accurate he further decreed that all the years which are multiples of 4000 shall not be leap years. Thus 2000 and 3000 will be leap years; 4000 will not be.

Such far-sightedness is of slender concern to those of us alive today; but one wonders if, in far-off electronic days, unimaginably more technological than our own, children will still be parroting the appalling rhyme of 'February hath twenty-eight alone, and all the rest have thirty-one'. Surely the Julian calendar with its Gregorian amendments will be replaced just as surely as farthings and halfpennies have had to go. It is only humans, and not computers, who are fond of such quirkish legacies. However, even the seasons may have been destroyed by then, and their uncontrollability will have been controlled; but, for the time being, and for the rest of this book, they are of paramount importance.

2 The Terrestrial Situation

With considerable myopia the people who live in the temperate parts of the world tend to assume that their kind of season applies all over the planet, that the days shorten as winter approaches, that leaves fall, that the nights grow chilly, then grow cold, then freezing, then longer, then warmer, and hot once again while the new leaves bask in the summer sun. Temperate people should spare a thought for some Siberians, whose spring comes in June but whose permafrost, or frozen earth, never leaves them. Or for the tropical regions where seasons are more a matter of rainfall, of deluge followed by droughts, and where the wind may blow consistently from one direction for half the year, only to blow with equal consistency from a contrary direction for the second half. No one in the tropics speaks of autumn. No one in Europe speaks of the trade winds, the monsoon or the hurricane months. Seasons, in short, are no one thing, but they are all different manifestations of that most singular property of the planet Earth, namely that it revolves around the sun with its axis at an angle to the plane of its rotation.

Just as spring, summer, autumn and winter have astronomical definitions, plus subjective associations according to locality, so do those localities. The tropics are hot, and the arctic regions are cold, but both areas have been defined without any reference to their climates. Between the latitudes of the tropic of Cancer (23° 27′ N) and the tropic of Capricorn (23° 27′ S) lies the tropical zone: only within this zone can the sun ever be directly overhead. It becomes directly over the northern tropic in June (when the sun is passing through the Cancer zodiac) and over the southern tropic in December (when the sun is passing through Capricorn).

Within the tropics, night follows day 365 times in the year, and always at the same sort of hour. One slightly boring feature of the tropics is this evenness of nightfall. Night also follows day 365 times a year within the temperate zones, but the farther a spot is from the equator the greater the possible inequality between day and night. In Scotland, although still within the temperate regions, the summer days stretch deep into the night, and the nights are fleeting, being quickly replaced by exceptionally early dawns. Within the Arctic circle, 66° 33′ N, or within the Antarctic circle, 66° 33′ S, days can be longer than twenty-four hours, and therefore less than 365 periods of daylight are interrupted by darkness every year. Both day

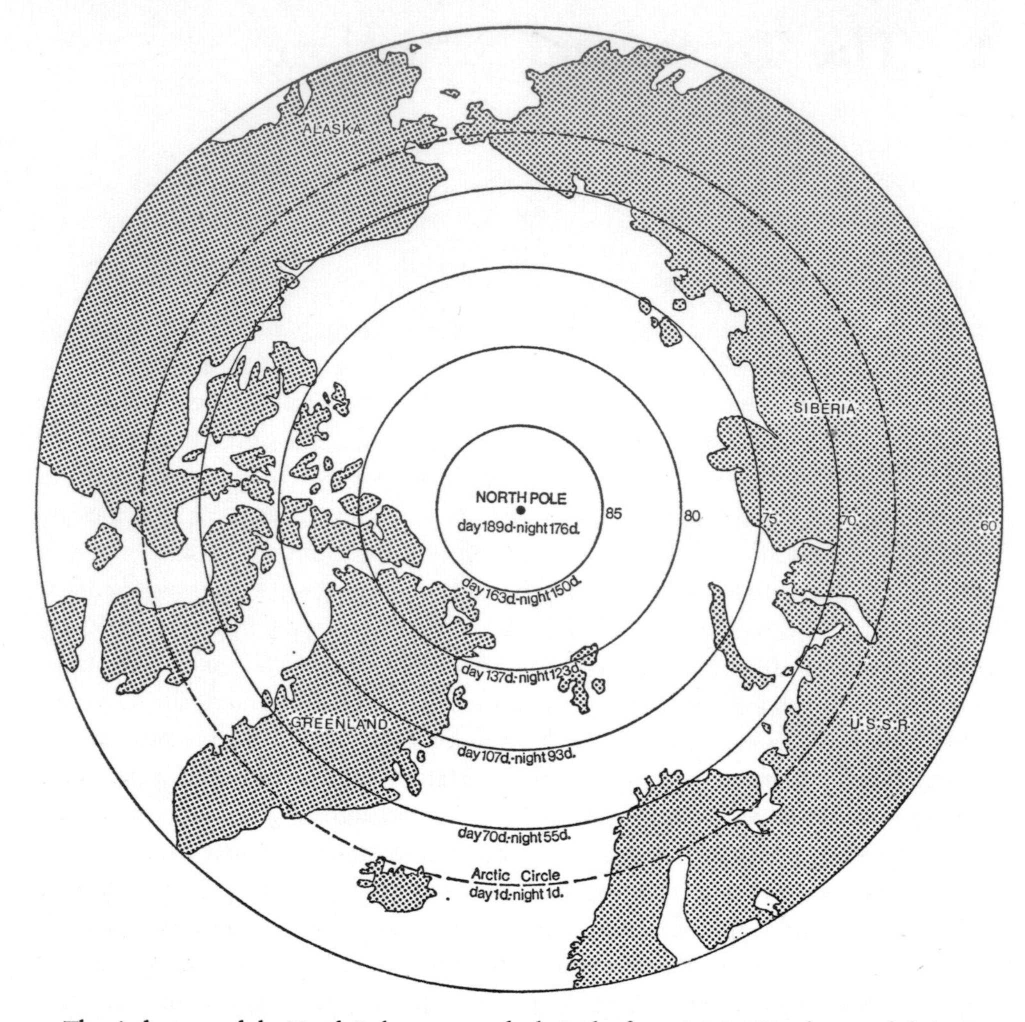

2. The circles around the North Pole represent the latitudes from 60°–85° North: at each latitude the lengths of continuous day and continuous night are given, one continuous day (d) being a period of 24 hours. In between the periods of continuous darkness (winter) and daylight (summer) these regions do experience the traditional night following day pattern. The longer periods of continuous daylight are due to the angle of the sun's rays on the surface of the earth.

and night can last for over twenty-four hours; but this fact is immediately associated with the problem which confuses the real astronomical situation.

In the same sort of way that a fish appears to be swimming nearer the water's surface than is actually the fact, so does refraction mean that the sun appears higher in the sky than is actually the case. The fault lies in our atmosphere which upsets the optics by bending the light waves. Refraction is at its greatest when the sun is near the horizon: the sun can still be seen even when it has gone below the horizon and would, without the intervening atmosphere, be invisible. For the two polar circles this means that they receive more sunlight than, geometrically speaking, they ought to. Not only do they receive sunlight after the sun has actually set, but they receive it again before it has actually risen. Therefore both poles are better off from the point of view of sunlight than they would be without the kindly beneficence of refraction.

At the extreme north pole, for example, the single night of winter lasts for 176 days, whereas the continuous daylight of summer lasts for 189 days. At 80° North, the latitude of northern Spitzbergen, the remorselessness of the wintry night is less eternal, and the perpetual daylight of summer is shorter, but the unbroken summer is still longer than the unbroken winter by 163 days to 150 days. (The remaining fifty-two days in the year have night follow day in the traditional manner once every twenty-four hours.) At latitude 70° North, a line which cuts through Lapland and much of the northern land mass of Siberia, the continual daylight of summer lasts for seventy days, and the steady darkness of winter lasts for fifty-five days. The remaining 240 days in the year include both darkness and light every twenty-four hours. Just 3½° farther south, and on the latitude of the Arctic circle, there is no longer even a short period either of perpetual darkness or of perpetual daylight: instead neither any day nor any night lasts for as much as twenty-four hours, and every year contains both 365 daytime periods, and 365 periods of night.

However, there is a further complexity. A summer's night even to the south of the Arctic circle bears little similarity to a tropical night even many degrees to the north or south of the equator. Not only is the Arctic summer night a relatively fleeting affair, but it never becomes completely dark, as happens so rapidly in the tropics. The reason is the angle at which the sun sets. The sun in the tropics sets nearly vertically whereas the sun near the Arctic circle slinks along almost parallel to the horizon. At dawn the tropical sun shoots straight up again (as with Kipling's thunderous dawn) but the sub-Arctic sun is exactly contrary; it reappears to slide along nearly parallel to the Earth's surface.

This extremely shallow dive of the northern sun, and its equally shallow reappearance, suggests correctly that the sun is never far below the surface. And whenever far below the surface it still has the power to repel total darkness. Obviously it shines brightly on the Earth when it is visible. Less obviously it still has a brightening effect when invisible from a certain point, but only just. In fact, it has power to repel total darkness when not more than 18° below the horizon. This means that all areas within 18° of the Arctic or Antarctic circles never achieve absolute blackness during their high summer nights. As all of Britain, of Holland, of Belgium, almost all of Canada, and places like Paris, Prague, Warsaw and Moscow are to the north of latitude 49° N (i.e. 67°–18°), they all achieve a state in high summer akin to the eternal twilight of the Arctic nights. The sun's shallow setting prevents the summer nights from having the inky blackness either of a wintry night or of a tropical night. The same goes for anywhere south of latitude 49° in the southern hemisphere; but, save for the actual continent of Antarctica, there is practically no land south of that parallel. (It is small wonder that most scientific facts, except in books on oceanography, relate to the northern hemisphere. The prejudice is justified not only because most people have lived there, do live there and will live there, but because so little land is south of the equator. Even three-quarters of Africa lies north of it.)

These are some of the basic astronomical points which both impinge upon the seasons and create them. There are others, such as the fact that the Earth precesses. Our planet is an imperfect sphere (as might be imagined, although it was only the first artificial satellites which gave really precise information about the Earth's slightly pear-shaped figure). One result of its own momentum, and its imperfect form being subjected to the attractions of the neighbouring moon, sun and planets, is that the Earth has a precessional movement. This means that its axis turns round in space; its orientation changes. Every 26,000 years the Earth's axis is, so to speak, back where it started, having traced out a cone in the meantime. Consequently Polaris, the northern pole star, does not always possess the navigational pre-eminence it is at present enjoying. In 5,000 years' time the celestial pole will be near Alpha Cephei. It was near Alpha Draconis 5,000 years ago, and the Southern Cross could then be seen from England. (In fact, Polaris, the currently favoured pole star, is not even at the celestial pole today, but about 1° from it. As the full moon, for example, subtends an angle of about half a degree as seen from the Earth, Polaris is obviously quite inaccurate, although it is becoming more accurate year by year. In AD 2100 it will be as accurate as it ever will be – less than half a degree from the true celestial pole. Thereafter it will recede, and will not again reach its position of maximum accuracy until AD 28100.)

The neatness and exactness of each day lasting twenty-four hours is also questionable. As everyone knows, the rotation of the Earth causes an apparent movement of the stars at night. Those near the pole star appear to go round it, while

Temperature and rainfall ABOVE: Climatic zones are determined by the angle and rotation of the earth round the sun, and roughly follow lines of latitude. This general pattern is, however, modified by local factors such as sea, altitude and atmosphere, and by rainfall (BELOW). OVERLEAF: The moon, as well as the sun, plays a part in determining climatic conditions on the earth, seen here from the moon's surface.

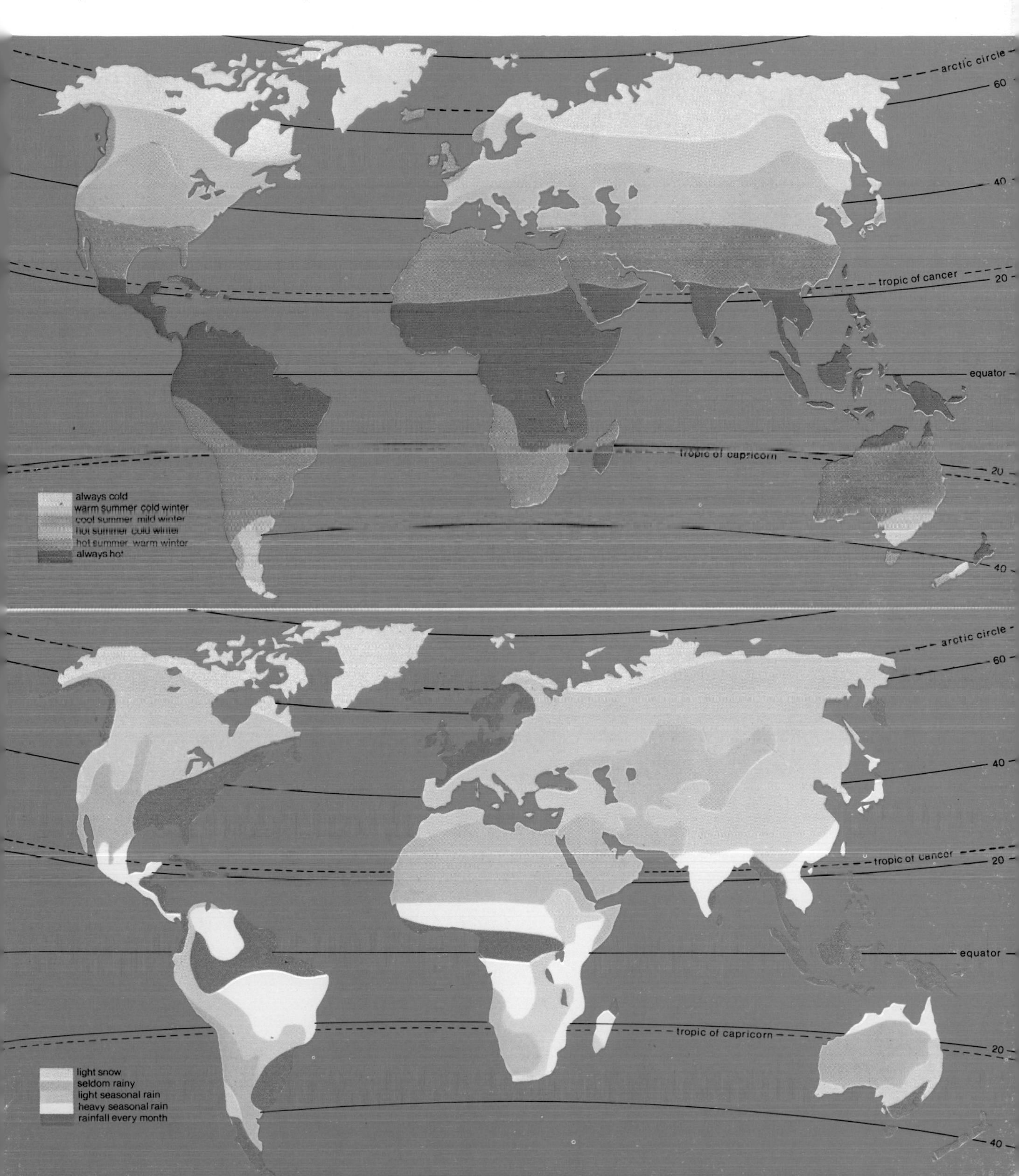

The angle of the earth's rotational axis to the sun determines the length of day and night. Within the Arctic and Antarctic circles continuous daylight (or darkness) can last from twenty-four hours up to almost six months.
RIGHT: The midnight sun in Siberia.
BELOW: The snow-stricken settlement of Barrow Point in Alaska.

The effects of rain (or the lack of it) can be dramatic. The two faces of Mount Kilimanjaro: rainfall is precipitated on one side of the mountain (ABOVE) while the other (RIGHT) remains dry.

LEFT: The degree of humidity can also affect the physical characteristics of animals. The lung fish, found in Africa and Australia, developed a functioning lung to breathe in oxygen from the air when its environment became drier.

Seasonal behaviour in animals Mating is the most obvious form of animal behaviour affected by seasonal change.
LEFT: A hedge-sparrow feeding a young cuckoo in its nest in spring.
ABOVE: The ferret begins its mating period in March (in the northern hemisphere) when the increasing length of daylight stimulates the appropriate hormones.
BOTTOM LEFT: The common frog spawns in spring.
BOTTOM RIGHT: Newly emerged tortoiseshell butterflies pumping up their wings.

OVERLEAF: The sun is the greatest single factor in climate. A solar flare (shown here) can affect the earth's atmosphere.

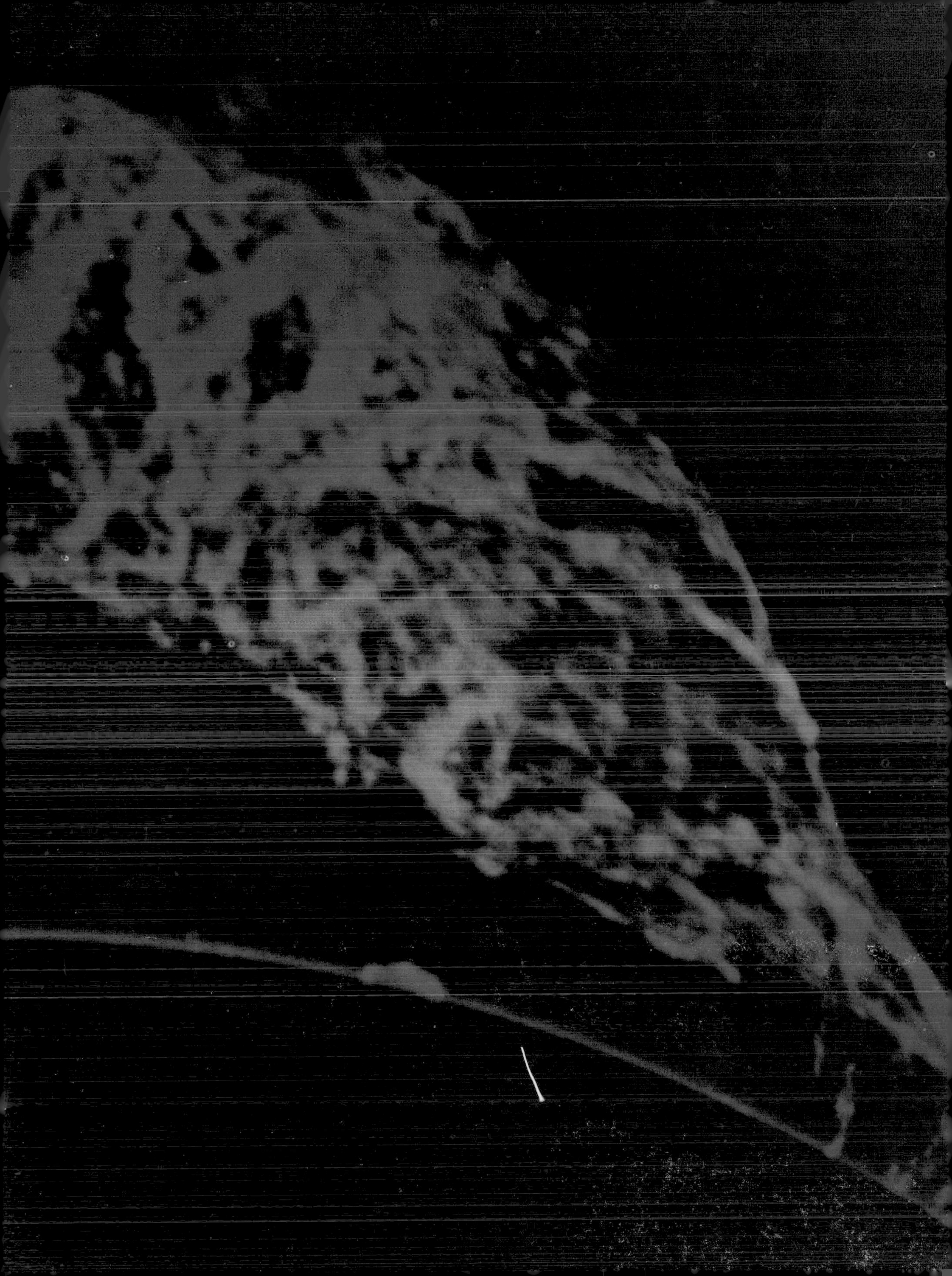

Seasonal variety Seasonal patterns vary greatly in different parts of the earth; the pattern of spring, summer, autumn, winter is by no means universal.
ABOVE: Spring in Siberia chiefly means the advent of the midnight sun. This multiple-exposure photograph taken at midnight shows the path of the sun just above the horizon.
LEFT AND FAR LEFT: In contrast to the frozen wastes of the Arctic, these desert areas are devastated by all-year round heat from the sun, combined with equally constant lack of rainfall.
RIGHT: Constant heat and rainfall produce the luxuriant vegetation of this Hawaiian rain-forest, which contains as many plant species in two square miles as the whole of the British Isles.

PREVIOUS PAGE: A solar flare.

Rhythmic change LEFT: Rhythmic change affects the geophysical environment as well as climate and animal behaviour. These rock formations, known as 'Goosenecks', are the result of a continued process of gradual erosion.
ABOVE: Tides provide the most obvious example of rhythmic change: every point in the sea has two tides a day, caused by the same gravitational pull of the moon that keeps the Mariner space probe (BELOW) in orbit round the moon.

those near the northern and southern horizons appear to travel parallel to these horizons. If a star is timed from one point in the sky until it has returned to that point on the following night it will have taken less than twenty-four hours. In fact, it will be back again at the starting place in 3 minutes 59·909 seconds less than twenty-four hours, and this means that the sidereal day, a time relative not to our sun but to our heavens, lasts for 23 hours, 56 minutes, and 4·091 seconds.

As a further complexity, the daylength relative to the sun, or to the heavens, or to anything, is changing. At first, when the extremely modest alterations were detected, it was assumed that the moon must be accelerating. This basic subjective fault is entirely as understandable as the presumption that one's own train has started to move when another glides out of the station. Now it is known that it is the Earth which is slowing down, and gradually its days are becoming longer. They are not doing so with any evenness, for plenty of other influences are at work, but within the next ten million years or so a whole day will have been lost. There will then be one fewer day per year than might be expected were daylengths to continue at the present level.

Most of us belong unofficially and surreptitiously (or so one has reason for belief) to the Flat Earth Society. It is far easier going about our business without being constantly provoked into concern that our planet is a ball hurtling through space, as is everything else. Consequently, it is easier to hold on to a few straightforward truths, that it takes twenty-four hours for a day to run, and 365 and a bit days for a year to occur, and the pole star is the pole star. If such simple precision is temporarily forgotten, and if our bizarre situation in the bottomless void of space is momentarily brought to mind, it becomes clearer that everything is changing all the while, that nothing will ever be quite the same again, and that straightforwardness is quite out of place in any universe, expanding or otherwise.

Such constancy as does exist is possibly more remarkable than the relatively minor inconsistencies. *Time* magazine once paraphrased our situation nicely when it postulated a citizen sitting quietly in his chair while both were, in fact, turning a gigantic somersault once every twenty-four hours, *and* circling at a speed of 700 miles an hour (assuming the chair plus occupant is on latitude 45°), *and* orbiting round the sun at eighteen miles a second (because the annual peregrination round the sun covers 580 million miles), *and* travelling in the direction of the star Vega, together with all our solar system, at a speed of twelve miles a second, *and* also gyrating around the centre of the Milky Way galaxy (of which the solar system is a minute part) at 170 miles a second.

Such a bewilderment of movement, and such an unimaginable assortment of speeds, make it entirely comprehensible that, unofficial flat-earthers as we are, we almost totally disregard or forget the complex machinery of astronomy. Of course the man and his chair are stationary. Of course they are going nowhere. How can one possibly visualize anyone revolving, orbiting, travelling and gyrating over immense distances, and all at the same time?

Despite such rational thought, these things are happening, and one way and another they combine to give us our nights and our days, our seasons and our years. There is no stationary armchair. There are only celestial bodies, and gravitational attraction, and the basic laws of the universe. The enormity of the system and the apparent eternity of its time-scales dwarf our own puny existence. So there is great difficulty, if not an impossibility, in being properly aware of the mega-galactic system in which we live. We know of our days, and of our seasons; but we disregard the convolutions which have caused them. We also consider climatic irregularities as being somewhat uncalled-for, and all regularity is believed entirely reasonable. We give scarcely a thought to the pulls and counter-pulls which keep us where we are and create such steady cyclical behaviour as does exist. Such reasoning, and the difficulty of comprehending the current situation, make it all the more staggering that astronomy was such an early science.

Admittedly the ancients got most of it wrong, and persisted with extraordinary explanations of Earth's existence and position, but a couple of millennia ago the knowledge of the heavens was remarkable. Ptolemy, the second-century Alexandrian, may have thought the planets revolved around the Earth, but his famous catalogue included 1,028 stars, all listed according to their magnitudes. The ancient Greeks and Egyptians attempted to predict eclipses, to weigh the Earth (both unthinkable tasks for the average well-educated citizen of today) and had an awareness of the heavens – judging by the written concentration on the subject – which puts to shame all those of us today who can pick out the Plough, the pole star, the Pleiades and little else. It was Copernicus in the sixteenth century who did most to put things right, who corrected Ptolemy, and who humiliated the Earth by insisting that it and all the planets went around the sun. It was Galileo who put things more right a century later when he had the unfair advantage of being able to see through a telescope. In short, the heavens have been the subject of intelligent scientific debate for thousands of years, and the planetary system has been understood for several hundred years, but little of this basic comprehension has become a part of us. Instead we talk of days and years with little innate comprehension of the whirling world which lies behind them.

Therefore, before discussing the seasonal situation here on this planet, it is best to think first of a planet without seasons, of one without days and nights, for such a planet is perfectly possible. Not only would it always revolve round the sun with the same side facing the sun, just as our own moon always faces the Earth unilaterally, but its axis would be vertical to the plane of its circular orbit. Provided this imaginary planet revolved once on its axis throughout each orbit, and therefore kept a steadfastly one-sided gaze towards its sun, there would be no succession of nights and days; and, provided each spot on the planet was always at the same angle to the sun, there would be no seasonal change. Each place would always be the same, either always hot, or always cold, or always nothing in particular. Much like a cave, which tends to know no seasons, or the controlled comfort of a chromium-finished hotel, both the daily round and annual cycle would have an indistinguishable sameness to them. The stimulating inconstancy of each day's progression, and of the seasons, would be totally lacking. It would be a situation quite unknown to us, and it would also have quite a different kind of life.

Evolution, one assumes, must have been profoundly affected by the perpetual motion of days and years. Imagine that hotel again. Imagine within it some well-nurtured inhabitant, highly accustomed to a thermostatic environment and a clockwork regulation of routine, who then sets forth for some unaccountable reason either into the depths of winter or the height of summer. Imagine, for the sake of this analogy, the adventurer hopelessly attempting to withstand the novel onslaught of cold or of heat. He would perish, swiftly, from the savagery of his changed environment. Customary lack of change is no preparation for an abrupt alteration in any situation, and evolution has always favoured the adaptable animal, the creature that does well and whose descendants continue to do well even when the environment changes. Living like a fish is one ability; living like a fish but producing modified descendants that can breathe air should the lake dry up, is quite another.

Consequently it can be argued that the day/night routine and the summer/winter shift have assisted the divergence and speed of evolution. Animals and plants have had to withstand fluctuating daylengths and erratic temperatures in the temperate and polar regions, and marked variations in dryness and wetness in the tropics. Certainly no zone is comparable either to cave or hotel. Nowhere is the same, day in, day out. Therefore, because animals and plants suited to a particular environment can readily adapt when that environment suffers its cyclical changes, they can also be expected to adapt should their world linger at one or other extreme, should the summers wax and the winters wane, should the nights grow colder or longer or windier or wetter.

It is presumed that, when change has come, most life accustomed to the previous situation does not thrive so well. On the other hand some of its forms, better adjusted to change, may survive even better, and such improved survival is the basis of evolution. In any case a species adjusted to a changing world is, according to this elementary argument, more likely to withstand some further change than another species adjusted to no change whatsoever. The farm labourer accustomed to working in all weathers is in better shape, one assumes, to overcome a further onslaught than the hotel guest accustomed to one most precise and finite world. (American hotels can seem to belie this comparison by keeping their rooms cooler in summer than in winter, thus inducing seasonal change, albeit contrarily. However, any guests unwise enough to leave will then find the summer heat yet more searing and the winter cold yet more penetrating, and so will suffer accordingly.)

The Earth's rotation on its axis must have been similar throughout the many hundred million years that life has been evolving. Such a quantity of momentum is not something which could readily be affected, at least to any marked degree, by the planets, the sun and the moon. The spin was certainly somewhat faster when life began, and therefore the days were shorter, but the general situation was probably similar. However this similarity of rotation, and the similarity of orbiting round the sun, and a general sameness in the behaviour of our fellow members of the solar system, by no means implies that the Earth's physical environment has been without significant change. Instead it has been perpetually changeable.

The world's map, today neatly delineated into the familiar continents, has certainly not been constant. In fact, the current situation follows from earlier continental situations practically unrecognizable in their outlines. Almost certainly the New World was once attached to the Old, with the great prominence of Brazil fitting into the equally great indentation of West Africa's Gulf of Guinea. The United States were previously disunited with a great sea in the area where the Rocky mountains now stand, and the creation of such new mountains, plus the levelling of old ones, has been an integral part of the eternal change.

Generally speaking, all the large mountain ranges of today, like the Alps, the Himalayas, the Rockies, and the Andes, are new mountains. The earlier and older ranges, like the Scottish hills, are worn-down stumps of their former glory. The new peaks give evidence of their previous levels by possessing fossil forms, often of sea-living creatures, in their lofty strata. At 20,000 ft in the Himalayas, for instance, it is possible to climb past marine fossils, and the same is true for the very top of some of the European Alps and the Canadian Rockies. As these fossils were laid down in Eocene times at the beginning of the Tertiary period any creature (for man was still absent) then wishing to climb the world's highest mountain would have had to look elsewhere than the Himalayas. Neither they nor Everest had yet been born.

Africa has been the most static continent over the ages. Apart from its Atlas region in the north and the Cape area in the south, the bulk of the continent has kept itself above water since the beginning of the Palaeozoic. Madagascar is thought to have budded off sometime to travel northwards as well as eastwards from the continent, but the shape of the continent has stayed remarkably uniform. So too has the central mass of Australia.

All changes to continents have inevitably had profound effects upon climate, and the present situation shows how the new mountains must have altered events. As the monsoon, for example, travels up the Ganges it drops its rain before reaching the Himalayas, and has none left for the deserts on the other side. The Rocky mountains (assumed to include the Sierra and the Cascades) experience prevailing westerly winds in the region of California, Oregon and the State of Washington. Consequently the western slopes of the westernmost ranges are well watered, but the eastern face looks down upon the driest area in the United States. Conversely, but following the same rules, the Amazon basin is deluged in rain from an easterly wind, but west of the Andes there are deserts. One desert area in Chile has never experienced rain in the 400 years since Western man has been in the area.

Quite apart from the one-sided results of mountain ranges, the continents themselves affect the weather. In general, places near the sea have a more stable climate than places far inland. If you travel around the world on the northern 40° parallel (and start at the Pacific) you pass through the temperate western seaboard of the United States, then the extremely hot and extremely cold American prairies, and reach the equally rigorous climate of New York. Then you pass over the Atlantic to the balmy lands of Portugal and southern Italy, to the increasingly bleak world of northern Turkey and western Turkestan, to the desert of Tashkent, the extremes of China, the temperate islands of Japan, and across the Pacific to the Californian coastal region. Even on any one day it is a latitude of considerable divergence, and almost all latitudes are similarly varied. All Britons live north of almost all Canadians; yet the harshness of the Canadian winter is out of all proportion to the gentle British one. Edinburgh is farther north than Moscow. The Shetland Islands are at 60° N; so is Cape Farewell on Greenland.

Also land itself is no one entity. Its altitude above sea-level is important, quite apart from the available rainfall. A generalization is that temperature falls 3° F (1·6° C) for every 1,000 ft rise, but local conditions can enhance this drop. (It is easy to experience this phenomenon when leaving Capetown. Within three hours, and after snaking uphill, shirt-sleeve weather can give way to temperatures demanding every muffler, glove and stitch of extra clothing, although the general level of the Karroo is only some 3,000 ft above sea-level and there ought to be no need to behave like an Eskimo. There must be local conditions aggravating the situation, such as a cold wind from which Capetown is protected.) The first visitors to bring back reports of a snow-capped peak in East Africa lying almost on the equator were scoffed at by the armchair savants; yet Mt Kilimanjaro, 19,324 feet high, visibly demonstrates a temperature drop of 19 × 3° F (19 × 1·6° C), or 57° F (30° C).

The steaming Congo and Amazon basins, the heat of Sumatra and Borneo, are all on the same latitude as the snows of Kilimanjaro.

All such facts should be adding to an appreciation of an inconstant environmental situation despite the constant spin of the Earth. There would be inconstancy even if the land masses stayed constant, but they do not. Mountains are built up, and then are washed away. Continents bulge upwards, then sink beneath the waves. Deserts bloom, then flourish, then dry up. Lakes appear, then vanish. The world of living things has to adjust, to move, or to perish. The actual causes for such cataclysmic changes may be most trifling; a little more wind here bringing less rain there, or a little rise here, a sinking there, and a whole lake system disappears. The loss of a lake may itself cause other changes, quite apart from the devastation to everything depending on that lake, that water, those streams, that mud, those banks, those tributaries and floods. Causes can be modest, and results can be severe, but rarely can the effects be so harsh as with the ebbing and flowing of the Arctic ice-sheet.

One of the most extraordinary facts about the last so-called Ice Age is not that it happened in some distant past, like the Jurassic and its dinosaurs, but that it occurred so recently. A mere 20,000 years ago, so recently that *Homo sapiens* was already an artist, practically the whole of Canada and much of the United States was covered by a colossal sheet of ice. It reached as far south as today's Ohio and Missouri rivers, and the site of St Louis (38½° N, the same latitude as Athens, Messina and southern Spain) must have looked as parts of Greenland look today. When and where the snows of winter melted in summertime there would remain the towering and massive presence of the ice which did not melt, and in central Canada this frozen sheet was probably two miles thick. Quite apart from all its more blatant effects in transforming fertile lands into a frozen waste, the huge weight of the ice – not inconsiderable when 10,000 feet thick – depressed the earth it was resting on. Marine terraces, in which shells have been stacked on differing balcony layers, give mute evidence of changing sea-levels, and show how the land has risen again and again as the ice has melted, perhaps by two inches a year for a thousand years. The shores are still rising quite rapidly in the Baltic and the Gulf of Bothnia. Consequently rivers change their course as the land itself changes; lakes dry up or form; islands vanish or reappear.

Britain suffered from the Quaternary ice ages but, due to its happy position at the receiving end of the Gulf Stream, far less than its latitude would suggest. Scotland and Ireland were covered, but the southernmost part of England escaped the searing efforts of the ice which, in America, had transported boulders from the region of Hudson Bay to the slopes of the Rocky mountains. The ability of ice to transport rocks and earth leads to its ability to form moraines, or accumulations of such material. Many lakes, such as Garda and Como in northern Italy and Lake Constance (all at a much more southerly latitude than Gulf Stream Britain) were formed by moraine deposits blocking the exits of the water melting from the ice.

In the history of the world, and not just of the very recent and Quaternary period, ice-sheets have existed in all sorts of seemingly improbable places. In the Permian glaciation there was ice in the eastern Congo actually on the equator, and most of Africa south of 20° S was ice-bound. So too was India, most of Brazil and large areas of south and west Australia. Perhaps the Permian, being at the end of the Palaeozoic era and 400 million years ago, is far too remote to be of great concern today, but the last ice age period was only the other day. More important still the length of time separating us from this last great ice encroachment is less than the time which separated the various ice ages of this period. Are we, therefore, likely to be smothered with all our works beneath another great sheet of ice?

In short, the climatic consistency which we enjoy, and which is seemingly so permanent a feature, is a most delicate balance between various forms of destruction. The mean annual world temperature is 58° F (14·4° C), slightly less than is comfortable for the average human being. Should it fall, for whatever reason, or for whatever bewildering complex of interacting reasons, the ice would inevitably advance. It has been calculated that a drop to a world mean of 54° F (12° C) would deluge us in ice as on the last occasion. On the other hand, with so much water currently locked up above sea-level in the form of ice, notably in Greenland and Antarctica, any increased warmth in the world would cause a melting of that ice, a rise in sea-level, and a flooding of all the land at present no more than 100 ft above sea-level. This would include, of course, every existing port.

Therefore, living between the devil of ice and the deep blue sea becoming yet deeper, it is small wonder that major plans to alter the Earth's surface, such as damming the Bering Straits or blocking up those of Gibraltar, will receive scant sympathy until the world is more certain of all possible after-effects. We balance, as West African editorials used to say in other contexts, on the twin horns of a double dilemma, wanting neither more heat nor more cold, neither an increase of ice nor of water. For the time being, despite the curious plethora of climates and conditions, we will make do with the devil we know.

To sum up. The past has been a time of permanent change, both in the sense that each change has lasted for a long time and that each change has inevitably led, in the end, to some other situation. The polar regions are now ice-bound; they used not to be. The deserts are now dry, but they have been lakes, or sea-beds or even mountains in the past. However, no creature living on an ice-sheet or in a desert can reasonably expect such momentous changes, for lifetimes of individuals or even of species are of quite a different order to the longevity of geological ages.

Nevertheless, during those static times, when desert was desert, or forest was forest, there always were the nights and days, and always there were the seasons. These permanent cycles acted as forceful reminders that constancy was not preordained, that stability was not for ever. The cold nights and hot days, the wet times and dry times, were at least some sort of preparation for the time when the planet itself was due to change. These days and years did at least instil adaptability to change into living things, and such quality must have been a vital asset when the whole environment altered, when the old order yielded to the new. The days and nights are of various kinds, the seasons are of various kinds, the types of adaptation are equally varied and will be discussed in later chapters; but the essential point of this introduction is that evolution happened largely in response to a changing world, and astronomy has always caused changes in the environment which are by no means irrelevant to evolution. Everything is interlinked – the days, nights, winters, summers, the shifting continents, the wavering ice-sheet – and all are intimately linked with the biological world existing today.

3 One Planetary Animal

A total change in approach is now necessary – a switch from astronomy and the climates to biology. Between the astronomical causes and the biological effects lies a gap which needs to be straddled to show the future range of this book. It is all very well to talk in general terms of an animal's adaptation to a seasonal environment, but this is far from explaining how such an adaptation takes place. Also many biological reactions to the daily and yearly rhythms are well known, such as migrations and hibernation, but being aware of these adaptations is a far cry from comprehending them. It is easy to remember that swallows leave Africa in the early months of the year, that some arrive in England in mid-April, that they then nest, lay eggs, hatch them and eventually fly back to Africa with their families; but it is a labour to think of describing all the actual mechanisms involved. How do the birds know when to leave Africa? What acts on what to produce the migration? What stimulates what to set which glands in action? Is it change in the African daylength, or an alteration in the humidity with the approaching rainy season? Or some awareness of day numbers, and therefore dates, or something quite unsuspected by man? It is all like that story of the Eskimos who were making thicker igloos one autumn. How can they possibly suspect it will be a hard winter, asked Americans from a nearby base. 'We saw you flying in more fuel than last year,' they replied.

The purpose of this third chapter is to bring the subject of adaptation down to the same level, the level of immediate practicability. Rather than discuss the whole multifarious biological world, with its innumerable reactions, it seems preferable at this stage to single out one relatively straightforward animal, and describe what is involved in the apparent simplicity of its annual cycle. What is the clockwork involved? What is it about spring that can possibly affect an ovary? The ferret is suitable, being neither overtly complex nor remarkably simple. It has a life-cycle which can be considered quite typical of the multiplicity of life-cycles existing in the higher echelons of the animal kingdom. Its story both diversifies and sums up the problem.

This animal is a semi-domesticated carnivorous mammal, somewhat bigger than a large rat, which is often used for frightening rabbits out of their burrows. It has a sleek, wiry and slightly ominous look, but some people are able to train them until they become, to all intents and purposes, pets. Their owners can even carry them, knowing how not to get bitten, in a coat-pocket.

To the scientist studying seasonal variations in behaviour, the ferret is not quite so tractable a beast. In England and other parts of the northern hemisphere ferrets breed from April to August. The first matings usually occur towards the end of March. Pregnancy lasts only seven weeks, and every mature female can raise two litters, each containing about six young at a time, during spring and summer. A period of sexual hibernation known as anoestrus then sets in, and throughout the autumn and winter the ferret is reproductively inactive.

This seasonal behaviour is associated with a curious physiological phenomenon. The females of most mammals which have been studied, including man, ovulate spontaneously. When they ovulate and release a ripe egg-cell from the ovaries, the act occurs without any assistance from the male; but ferrets belong to a minority of animals which ovulate only in response to the act of mating. Furthermore, when they are in the physiological state of readiness to mate, known as 'on heat', 'in heat', or 'in oestrus', and which occurs when egg-cells ripen in the ovaries, female ferrets advertise the fact by a swelling of skin below the base of the tail. This swelling develops in the spring at the beginning of the breeding season, and it disappears during pregnancy. If a ferret does not mate at all, the swelling persists throughout the summer, and only subsides when the non-breeding season begins in the autumn.

The skin swelling is due to the specific action of a chemical substance produced by the ovaries. The ovaries secrete this substance – called oestrogen – directly into the bloodstream, but the only skin which reacts to it is the small area below the base of the tail, the area around the external genitalia. The chemical nature of oestrogen is known, and it can now be made artificially. Therefore, although the swelling never occurs spontaneously in animals whose ovaries have been removed, it can be artificially induced by means of oestrogen injections.

The chain of events thus far is already quite lengthy. Ovulation is caused by mating. Mating is caused, or abetted (and the male too, has to be ready), by skin swelling. Skin swelling is caused by oestrogen. And oestrogen, in its turn, is not produced until the ovaries have been stimulated by a chemical from the pituitary gland, a structure hanging from the lower surface of the brain. Like the oestrogen produced by the ovaries, the ovary-stimulating chemical which the pituitary manufactures is also secreted directly into the bloodstream. Both belong to a class of substances called hormones which will be encountered in greater detail in later chapters.

Two more facts need to be mentioned before the problem of the ferret can be fairly outlined. The first is that if a normal anoestrous female ferret is kept in a room in which the electric light is switched on each day for a few hours after sunset, it comes on heat within one to two months. Let us assume that the experiment is started at the end of September. By this time all normal ferrets have completed their breeding and are sexually quiescent. Each day as darkness falls the electric light is kept on for six hours. Within eight weeks the skin-swelling will begin to appear, and by ten weeks it will have developed fully.

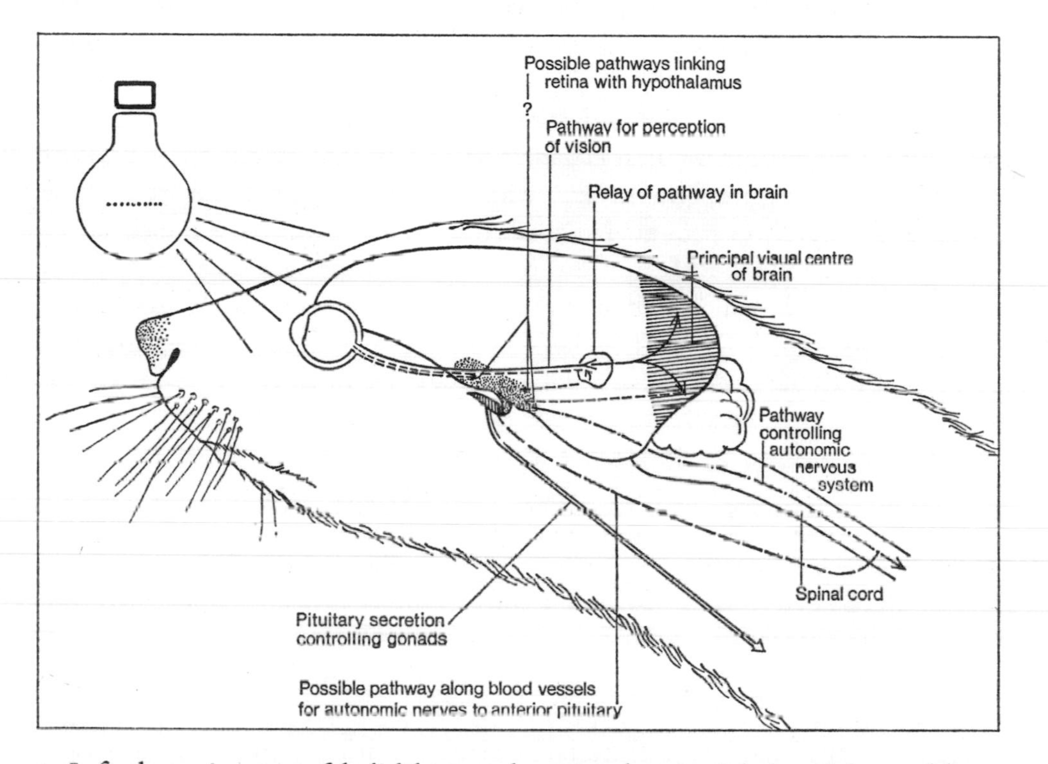

3. So far the precise nature of the link between the optic pathways and the hypothalamus of the ferret (see diagram) is not known, nor is it known how the hypothalamus stimulates the anterior (or front) part of the pituitary gland; but it is known that the pituitary gland secretes some kind of hormone (or chemical) which activates the ferret's reproductive organs.

The second point is that nothing happens if the same thing is done to a blind ferret, whether it is naturally blind due to cataract or unnaturally blind because its optic nerves have been cut. The animal remains anoestrous. The obvious inference is that the electric light stimulates the ferret to come on heat not by any action on the skin but by way of the eyes.

Here, therefore, is the problem. The oestrous situation does not occur if the ovaries do not secrete oestrogen, and the ovaries do not secrete oestrogen if they have not been stimulated by the pituitary. An anoestrous ferret from which this gland has been removed does not come on heat when exposed to additional light during the winter, any more than a ferret whose ovaries have been removed. When light enters the eye, it stimulates the retina at the back of the eyeball, and impulses are set up in nerve fibres which leave the eye as the optic nerve. The only direction such nerve impulses can take along the optic nerve is inwards, and towards that part of the brain concerned with vision. There must clearly be some kind of link between the optic nerves and the pituitary because, as already mentioned, the ovaries cannot act until this gland has first liberated its ovary-stimulating hormone into the bloodstream.

Unfortunately (for any easy dissipation of the problem) the parts of the brain with which the optic nerve fibres connect, and which are concerned with visual perception, have no possible *direct* link with the pituitary gland. However, nerve tracts do connect them with the hypothalamus, and the pituitary is suspended by a stalk from the under-surface of this hypothalamus, the specialized part of the base of the brain in which a number of more or less distinct collections of nerve cells can be identified. Together the hypothalamic nerve cells act as a vital organizing centre for the autonomic, as contrasted with the voluntary, nervous system of the body.

One could assume, therefore, that the second phase of the chain reaction by which the ovaries are stimulated by light is the activation of nerve cells in the hypothalamus, being a secondary effect of the stimulation of the brain directly concerned with vision. The third phase would be the activation of the secretory cells in the pituitary gland by nerve impulses that are first fired along the nerve fibres of the hypothalamic cells down the stalk of the gland. The pituitary stalk does, in fact, consist of thousands of nerve fibres which connect the hypothalamus with the pituitary gland: but few, if any, of them – or so it is usually claimed – reach that part of the gland which secretes the hormone that stimulates the ovaries. How, then, does the stimulus of light work? How does the basic astronomical cause have its undeniably biological effect?

The current doctrine is that the nerve cells in the hypothalamus not only send nerve impulses down the pituitary stalk, but also secrete a chemical substance. This is presumed to be a kind of local hormone. It escapes from the hypothalamic nerve fibres near the start of the pituitary stalk, and filters into, or is picked up by, minute blood vessels which discharge into the blood vessels of the pituitary gland. The presumed hypothalamic secretion is, on this hypothesis, a specific hormone which controls the pituitary. Anyhow, whether all this presumption is eventually vindicated or not, it seems a far cry from the fact that one planet orbiting in space takes about twenty-four hours to rotate on its axis and about 365 days to complete each orbit.

The problem of the ferret is a fairly typical conundrum of biology as well as being germane to the central theme of this book. It reveals only too clearly that biology is not a simple science, like physics or chemistry. The constituent parts of a question in the physical sciences can be defined with reasonable assurance. For example, the chemist might wish to find out the structure of some substance, or to learn how to make a complicated molecule from simple molecules. Even when the problem is so difficult as to defy solution, the chemist is fairly certain of its exact nature. This is the exception rather than the rule in the study of living processes. The analysis of almost any one of them is at every turn beset by unknown factors, and the experimenter hardly ever knows what number of variables should be taken into account when he is trying to unravel a particular knot. Usually it is easier to disprove some theory in biology than to show which theory should be adopted. Hence, a major difficulty with the establishment of biological facts is that the biologist is always uncertain if he has taken note of all variables which may be involved and may be influencing the facts. And where facts fail, fancy comes in if the biologist is to satisfy his basic scientific urge to explain things.

Where fancy reigns, so does fashion. The hypothesis which has just been outlined, about the way the pituitary's ovary-stimulating properties are controlled, is now all the rage, even though its basis is so slender as to constitute little more than arbitrary speculation. The point which the story of the ferret is meant to illustrate at this stage is that the whole question of the way an animal fits or articulates the phases of its life-cycle to the sequence of the seasons is, on the one hand, so fascinating as to stimulate the most enthusiastic of scientific inquiries, and on the other, so difficult as to end in facile answers which, by challenging credulity, demand increasingly critical and searching study. The broad facts of seasonal behaviour are easy to describe. It is their explanation which is so difficult, and which constitutes so searching a problem.

The puzzle is one which applies to all living matter. The life of the plant is organized by the seasons, and plants do not have eyes. Butterflies are seasonal, and they do not possess a hypothalamus. Worms are seasonal, and they lack pituitary glands. Seasonal behaviour is clearly independent of any single physiological adaptation. Moreover, cyclic behaviour must have a physical or physiological basis in every species whose life is organized by the rhythm of the seasons.

Having posed the problem, and having described in fair detail one relatively simple animal's fairly complex cyclical behaviour and response, it is now necessary to return again to the seasons, to the root causes of seasonal activities. The seasons are imposed upon the planet Earth by its astronomical status in the heavens, and they impose upon the biological world. They are the cause. Therefore they and their cause must first be examined in more detail before it is possible to move on in greater detail to the multiplicity of those biological effects.

4 The Basic Rhythms

The energy that animates the living matter of this planet Earth, every flower, tree, and living creature, is derived from the sun. Extinguish the sun, if such a thing can be imagined without the instant disappearance of its attendant planetary system, and such heat as lies within the core of the Earth would be insufficient to carry us through – what? – a week? – a month? – certainly some short finite period, after which death would supervene. The plants would go. The animals which live only, or predominantly, on green matter would go, and then those animals which prey on other animals, including ourselves.

The vital sun is a vast mass of incredibly hot incandescent gas, and its diameter is 100 times that of our globe. Its surface temperature is greater than 6000° C. Its interior is many times hotter and it is estimated that four million tons of solar matter are converted each second into radiation, a minute fraction of 1% of which reaches the planets, the remainder being dissipated in space. The total energy income from sunlight at the Earth's surface is about 110 million kilowatts. According to Professor A. R. Ubbelohde, of Imperial College, London, about a twentieth of this energy is converted into the energy of rainfall, and about one five-thousandth is converted into chemical energy and therefore saved up by vegetation. The world's coal and oil reserves are equal in energy to about three and a half days of sunshine, and we are happily consuming this store at the rate of two and a half minutes a year.

The energy output of the sun, and presumably of all stars, is not constant; therefore the amount received by the Earth is not constant. Every now and then comes a sudden outburst of activity, resulting in an increased rate of bombardment of the upper atmosphere of Earth by radiations which affect radio-transmission and which may result in magnetic storms. The outbursts can sometimes be seen in photographs of the edge of the sun's disc. Such an outburst occurred early in 1956, when a bubble of gas appeared on the eastern side of the sun, and in the space of minutes grew until its cap was 20,000 miles across – practically the circumference of the Earth. This cap then shot off at great speed into space, and remained visible for a distance of 200,000 miles. The force of this solar explosion was likened to that of 100 million hydrogen bombs, and the force behind its final burst was said to be 1,000 times the pull of the Earth's gravity.

It is all but impossible to appreciate these magnitudes, or those that are involved in any direction of the universe around us. The nearest star, other than the sun itself, is Proxima Centauri. It is four light years away from our planet, or four times the distance that light travels in a year, namely four times six million million miles. But what on earth is six million million miles in the light of ordinary experience? The greatest distance apart of any two points in the British Isles is little more than 600 miles. The distance between London and New York is 3,500 miles. The circumference of the Earth is only 24,000 miles. And 24,000 miles goes 250 million times into a light year. The Concorde is designed to fly over the Atlantic carrying its passengers at 1,500 m.p.h. and would therefore take nearly half a million years to cover the distance measured by one light year. At 1,500 m.p.h. it would take nearly two million years to reach the nearest star. At that speed it would take something like seven years' continuous flight even to cover the distance between the Earth and the sun.

None of this articulates with the world of everyday experience. But a slight sense of reality, or at least familiarity, can be imparted to some of the statements of magnitude by repetition, by scientific fiction, and by the man-made satellites. In fact the satellites, which have to be projected sufficiently fast and sufficiently far so that they go on encircling the Earth in an orbit of their own instead of falling back due to the gravitational pull, help greatly with an understanding of the universe. They orbit the Earth according to the same rules that keep the Earth revolving in orbit around the sun. And the sun, like the Earth, is rotating on its own axis, even if twenty-five times more slowly.

The other eight planets of our solar system move in different but similarly constant orbits, and – except for Pluto – in virtually the same plane as the one in which Earth moves. Two of them, Venus and Mercury, have their circuits closer to, and move much faster round, the sun; the Earth comes third; followed by Mars, fourth, with Pluto, the farthest planet, nearly 4,000 million miles from the sun. The distribution of the infinitesimal fraction of the sun's energy which reaches the planets depends primarily on their distance from the sun.

The amount of energy which beats down on Earth varies between day and night, and also with the seasons; and has always varied in this fashion. The way it all started is lost in the mists which shroud existing ideas about the origin of the universe. What is certain, is that there were days and nights as soon as there were planets, that there were seasons as soon as there were planets, and that there are seasons wherever there are planets. The planets and the seasons were also probably – and on our planet certainly – in existence before life began.

Since it is inconceivable that life as it is known on Earth, whether as viruses, bacteria, or human beings, could exist on the sun, it follows that from the moment matter took on these characteristics by which we define life, it was subjected to the diurnal, seasonal and longer-term variations in the solar energy which reach every part of the globe. The periodic phenomena to which the Earth is subject must have been imprinted on living matter from the start. The problem of the periodicity of the ferret's behaviour is thus only one extremely small part of the whole problem of the adaptation of living matter to periodic and irregular variations in energy input.

Facts about the possible existence of life on any other planet are scarce, to say the least, and will be discussed in the next chapter. However, there is no question about the fact that other planets experience a night and day, as well as seasonal variations. That is so because they are tilted to the plane of their orbits round the sun, and because they are rotating on their own axes, some faster than Earth, some much more slowly. While the Earth's sister planets exercise some gravitational attraction on each other and on the Earth, this is so slight relative to the gravitational pull of the sun on the whole family of planets that it has no practical effect on the Earth's daily rotation and annual journey, and therefore on the Earth's seasons.

On the other hand the moon, although very much smaller, has a much greater effect on the Earth's environment. It circumnavigates the Earth in an orbit which is slightly elliptical, and which passes a little less than a quarter of a million miles from us. The moon's diameter is only about a quarter, and its mass about one-eightieth that of the Earth.

The moon imposes a rhythm on things which affects life on its parent planet. Because the Earth is so close, relatively speaking, to the moon and so big in relation to it, the gravitational pull of the moon on that part of the globe which is closest to it is nearly 7% greater than the part farthest away. The Earth (notably the water which covers the Earth) is sufficiently elastic to yield to this force, and becomes pulled out along a line joining it to the moon. This pulling, added to the greater effect of the Earth's own rotation on its equator, causes it to be still less of a sphere than it actually is (its diameter is always slightly greater across the equator than across the poles) and more like a symmetrical ovoid. This distortion affects the oceans, which get heaped up for a few feet, with the subsequent movement of the sea on the shallower shores causing the conspicuous tides. As the moon continues on its circuit, the point at which it exercises its maximum pull passes, and the place on which it had just focused immediately begins to spring back, as it were, to its original shape. The high tide then recedes down the beaches once again.

The gravitational pull exercised by the moon is responsible for 70% of the force involved in raising the tides. The sun contributes the remaining 30%. Both when the moon is new (it then lies directly between us and the sun) and when it is full (it then lies farther away from the sun than we are) the moon and the sun combine

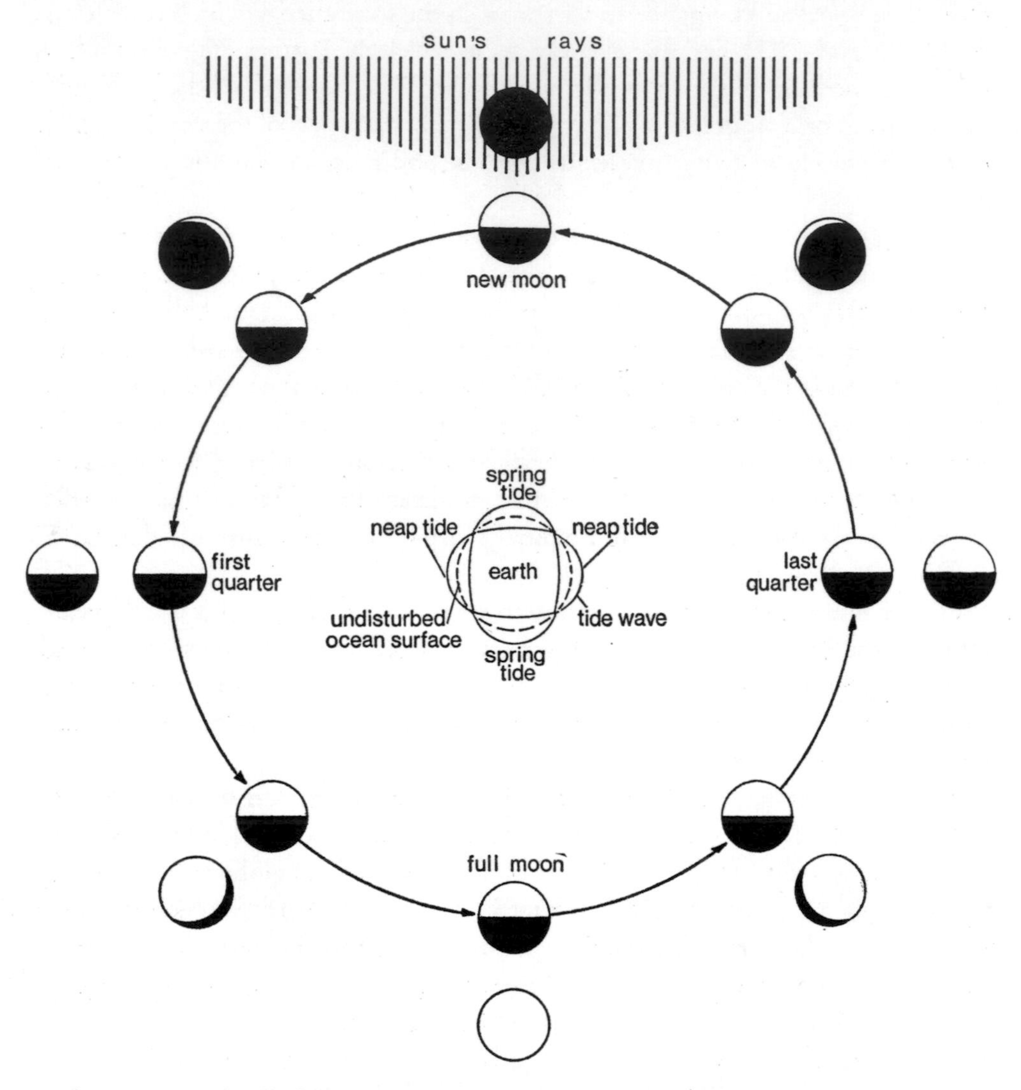

4. The gravitational pull of the moon is responsible for 70% of the force involved in raising the tides. The sun contributes the remaining 30%. When the moon is new and when it is is full, it lies in the same line as the sun, and, together, their gravitational pulls influence a high, or spring tide. At the moon's first and last quarters it is at right-angles to the sun's pull and the two forces tend to cancel each other out, producing a low or neap tide.

their gravitational pull on the Earth; as a result the coasts experience high spring tides. When the moon is in its first and last quarters, its pull is at right-angles to that of the sun and the two forces tend to interfere with one another, with the result that the high tides are then lower; they are the neap tides.

Either way, every point in the sea experiences two high tides each day, the first soon after the moon has focused its gravitational pull directly on that point, and the second, some twelve hours later, when the converse is true. The Earth has then rotated half-way round on its axis and the moon is then pulling in a direction opposite to the original line of pull. In between each high tide occurs a low tide, when the seas that have been piled up on the shore retreat again. The seas were on Earth before there was life on Earth. The tides, therefore, are a primeval rhythm, just as much as the alternation of days and nights and the flow of the seasons. All have imposed a ceaseless rhythm on living organisms from the moment life began.

The tides are regular, and so are the nights, and the rainy seasons, and the hot times, and the damp times. The ceaseless change is not just random change, but cyclical. It is difficult to think of any activity on Earth which does not have a cycle to it. The seasons were happening when the first sedimentary deposits were being laid down and, much like the annual rings of a tree, the years have been imprinted visibly on many rocks. Deposition, whether of animal debris or silt, is a cyclical business. The microscopic life of the sea flourishes in accordance with the changing ocean currents and with the seasons; the carcasses, resulting from each annual upsurge, rain down upon the sea-bed in mute and regular testimony to the annual cycle of events. The seasonal rain in the mountains brings down more earth in the rivers; and more silt is deposited year by year, season by season. The sedimentary rocks are built up year by year, layer by layer.

Such an eternity of regularity has built itself deeply into the biological world. Not only, as is well known, and as the ferret exemplifies, are the breeding cycles kept in tune with the changing year, but the seasonal metronome has pervaded living systems in many other ways. The so-called circadian rhythm (which necessarily has a chapter on its own) is an awareness of the daily cycle of events even when outside conditions, such as temperature and light, are kept constant. The internal rhythm pulses on, indicating the passage of time. Countless organisms have proved the existence of some endogenous pace-maker, a biological clock, which ticks away roughly in tune with the passing nights and days. Countless humans, flung abroad by the instant travel of today, acquire sufficient nourishment on their arrival, but their bodies stay tuned to the former pattern. They wake up repeatedly in the small hours, despite the darkness and the immediate need for five more hours of sleep. It is an irritating feature of an organism happily innured to one longitude and then unkindly transported to another.

Even circannian rhythms (of 'about a year') have been encountered in biology.

Some squirrels kept at constant temperature and with constant days and nights each twelve hours long have still reacted to the passing of the years. Their body weight has risen and fallen annually, they have exhibited hibernation behaviour at the appropriate time, and they have amply demonstrated the existence of a long-term yearly rhythm. It seems (more about this later) that external events, such as summer/winter and day/night, may on occasion not so much cause a rhythmic reaction but keep it in trim. Less like a watch-maker, and more like a watch-adjuster, the days and seasons have so imprinted their cycles on to living things that the deeply ingrained biological clocks are a result; the actual seasonal changes of each year merely regulate these existing mechanisms.

There are habitats which seem totally without seasons and without cycles (and these are described on page 167), but such unchanging spots are rare. Most forms of life suffer regular cycles of events; the desert is not always equally hot, nor the poles equally cold. Cycles are dominant and always have been. Animals and plants have responded, and have always done so. It is almost a law of planetary worlds that change should occur, and that it should recur cyclically. Life, once created, must change accordingly or perish. If there is life elsewhere it too must be subjected to the same laws.

5 Seasonal Life Elsewhere

The citizens of Earth are for ever in two minds about the existence of life in other places. They like to think they are unique and then, if extra-terrestrial life has to be, they like to think of that other life in a strictly terrestrial manner. It was easier back in those pre-Copernican days when the Earth was the centre of the universe, to assume that mankind was similarly important, a one-off product of creation. Now that our planet is known to be attached to a fairly insignificant star in a fairly insignificant position towards one edge of one galaxy out of millions of galaxies, our feeling of self-importance ought to have been pulverized. Instead, remarkably, it thrives, and it is still hard to realize that life probably does exist out there. When we finally accept this point it is harder still to imagine this life in any form but our own. It was for this reason, as Fred Hoyle has explained, that he wrote his book *The Black Cloud*, for the gaseous brain of that novel belonged to quite a different kind of creature. So too was whatever was happening, or had been happening, on the planet Jupiter in the film *2001*.

So far as is known, no extra-terrestrial intelligence has yet made itself known to us, despite the efforts of Project Ozme. That experiment, initiated and then concluded in the early euphoric days of the space age, was an attempt to discover intelligible sense or rhythm in the perpetual radio bombardment hurtling through space. Nothing was heard, the project was abandoned, and we are on our own again. We are leaving it to them to make themselves known to us. But if there is anybody out there, as the American physicist Edward Teller has remarked, 'then where is everybody?'

Various answers to Teller's question can be postulated. The first is that the distances are so big, and the number of planetary oases so small relative to the distance between each one, that space travel in any major fashion is not feasible for anyone, however advanced. Secondly, if actual travel is avoided, and messages are sent instead, the distances are still huge. A radio wave cannot travel faster than 186,000 miles a second, and therefore a reply even from the nearest solar system will take a long time. It would take some ten years to send a message and get an answer back from the star nearest to our solar system, and some 10,000 million years to and from the farthest stars we can detect. Thirdly, any other life is likely to be primeval, relative to our standards, or else enormously superior to us. It may be still in a primitive form, as in our own case for all but the last few decades of the several thousand million years of evolution, or it will be many technological millennia ahead of us. In the former case radio waves will not emanate from any hypothetical form of life which might exist, neither could it receive them. A coincidence of technical parity is totally improbable.

A fourth possible reason for the silence of other worlds is even more depressing. Scientists have proposed that *lambda* in this context should represent the length of time between a civilization developing the means for its self-destruction and the realization of that potential. Here on Earth we know full well that the means came into our hands with the manufacture, and use, of the first nuclear weapon in 1945. Within twenty years, at most, there was a sufficient stockpile to achieve the destruction of every living person on Earth, had this weaponry been detonated. *Lambda* for the planet Earth might therefore have been twenty years. In other words there had been some 4,000 million years while life was quietly maturing, then a few thousand while it was developing technologically, then a few decades of radio awareness, and then there could be silence. There might still be that silence.

If self-annihilation is a traditional pattern of progress this would explain the greater celestial silence. Failing that, and as a fifth reason, there is the entirely understandable possibility that more advanced civilizations might have learned the wisdom of not contacting other worlds. Colonial history on Earth has shown the existence of this other factor, which could be called *mu*. This is the length of time after a new place has been discovered before either the natives themselves grow restless or the colonists themselves wish to be independent. There is much to be said for staying at home, and for resisting the impulse to build empires. Both *lambda* and *mu* could explain why one asks 'where is everybody?'

Despite the silence, and there are reasons for it, the general opinion today is that there are 'people' out there. The Harvard biologist George Wald, for example, argued the case, although such an argument starts off with the apparently contrary point that planets like Earth must be rare. Life needs some warmth, not too much, not too little, and our own planet is excellently placed in its rare situation. Nevertheless the universe is so large that this kind of rare situation must actually be abundant. According to Wald there are at least 100,000 planets like the Earth in our own Milky Way galaxy. Unfortunately, this does not make them particularly accessible, because this single galaxy has a diameter of 100,000 light years, a figure which has to be multiplied by six million million for those who prefer to think in miles. The existence of 100,000 planets so relatively near in space does not make them visible either, in that no planet beyond our own solar system has ever been observed.

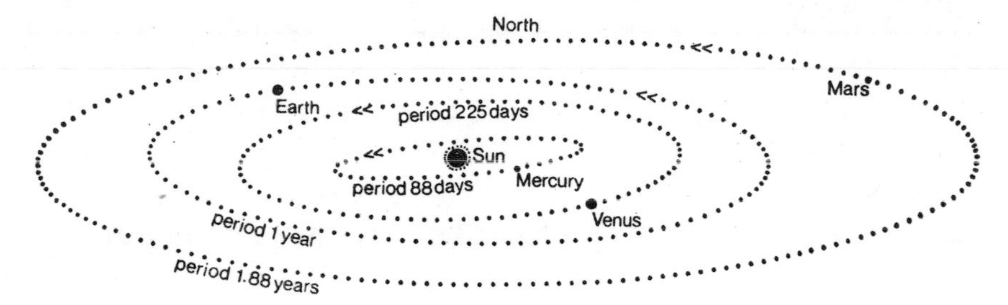

The orbits of the 4 inner planets

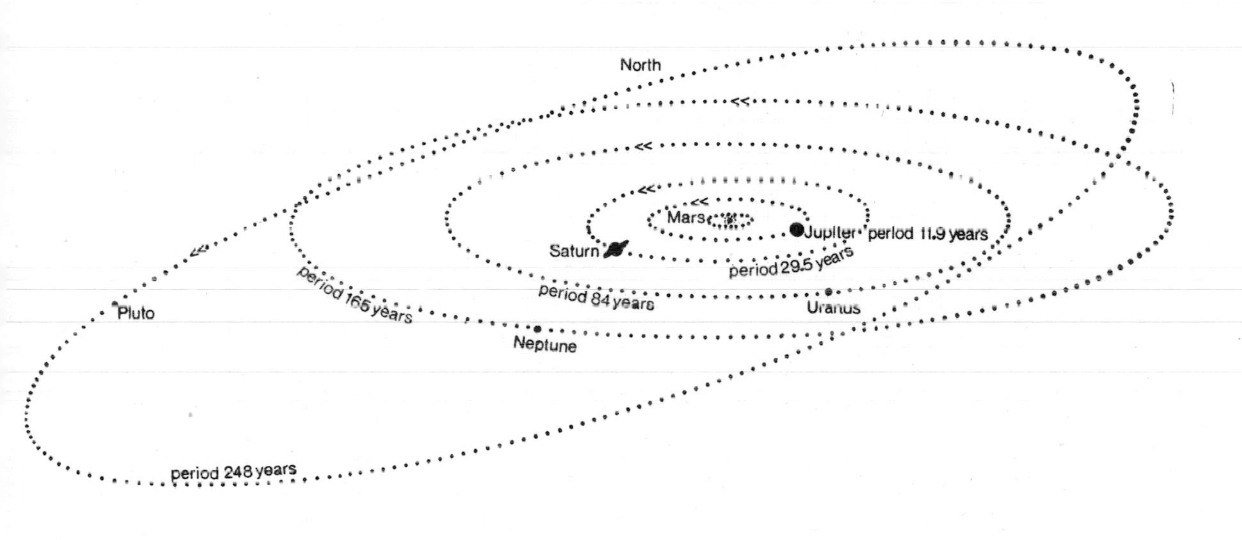

The orbits of the planets from Mars to Pluto
Scale about 20 times smaller than above

5. The diagrams show the orbits of the major planets around the sun. In the top diagram only the four minor planets are drawn, showing how the elliptical paths of both Mercury and Venus are described inside that of Earth's. The lower diagram (the scale is twenty times smaller) shows the elliptical paths of the outer planets. The periods of revolution of the various planets differ from approximately 88 days for Mercury to 248 years for Pluto.

Nevertheless they are presumed to exist, and over 100 million galaxies have so far come within range of earth-based telescopes. Multiply one galaxy's number of planets by the number of galaxies, and the resultant figure of ten million million is the number of planets like our own within the observable universe. Even if this is false by a wide margin, and even if only a fraction of that number are in fact suitable cauldrons for the generation of life, there are still very many planets with life on them.

Life, says Wald, is a 'cosmic event – so far as we know the most complex state or organization that matter has achieved in our cosmos. We are not alone in the universe, and do not bear alone the whole burden of life and what comes of it.' In this context the word life is used to describe what we know as life. Some argue that life has to be as we know it, that it has to be based on the carbon atom, as ours is. Others argue that one or two other atoms, such as silicon, would also serve. No one suggests that it can be so different as to conflict with the inescapable laws of the universe. Wald concludes:

> I am convinced that there can be no way of composing and constructing living organisms which is fundamentally different from the one we know . . . Wherever life is possible, given time, it should arise. It should then ramify into a wide array of forms, differing in detail from those we now observe (as did earlier organisms on the Earth) yet including many which should look familiar to us – perhaps even men.

In short, there are galaxies galore, there must be planets galore, and life must equally well be dotted all over the place. Wherever it exists matter must obey the same rules, for it is the same matter. Therefore, as here on Earth, living material can only exist under limited and specialized conditions. Knowledge and experience of our own planets can help to indicate more precisely these particular parameters. Even though our sun is suitably docile and well contained, and even though its planets are all old enough with surface decay being well advanced, life certainly does not exist on every one of this solar system's planets.

However, all planets, whether life-bearing or not, travel round an energy-giving sun, either in a fairly circular or highly elliptical orbit, and as it is the nature of heavenly bodies to revolve on their axes – even the sun does that – both seasonal rhythms of a sort and daily rhythms of a sort will occur. Such events are therefore not the sole prerogative of the planet on which we live: they are universal. Consequently whatsoever life does exist will be subject, to some degree, both to a daily and to an annual rhythm: there will be days and nights, and there will be seasons. The actual timing and extent of these events will surely be dissimilar to those existing on Earth, but extra-terrestrial life must still be subject to such daily and yearly cycles.

The planets Seasons, although rather longer ones, exist on Mars (RIGHT), whose rotation is not unlike that of the Earth. Some sort of life, if only bacteria or lichen, may exist on Mars, but seasons or life as we know it are out of the question on Venus (FAR RIGHT), which has a surface temperature of about 800°F (427°C). Somewhere in the picture below (although we cannot see it) there may be a planet on which life exists: it has been estimated that over ten billion planets similar to earth may be capable of sustaining life.

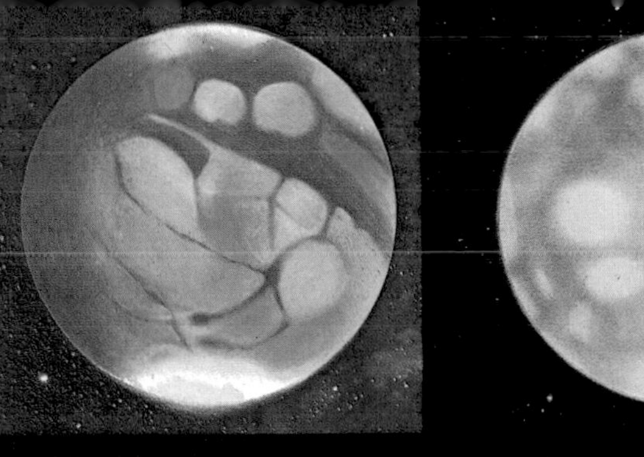

Prehistory In the several thousand million years since invertebrate life began, the earth has been subjected to immense geological and climatological change: ice sheets once covered Brazil; Africa and America were once joined; tropical fossils are found under central London. Fossils are our only source of knowledge for the prehistoric period.

LEFT: Fossilised vegetation of the Eocene period found in Bournemouth.

ABOVE: Fossilised shell, Carboniferous era, from Moscow.

BELOW: Salt crystals, Permian era, found in the Punjab.

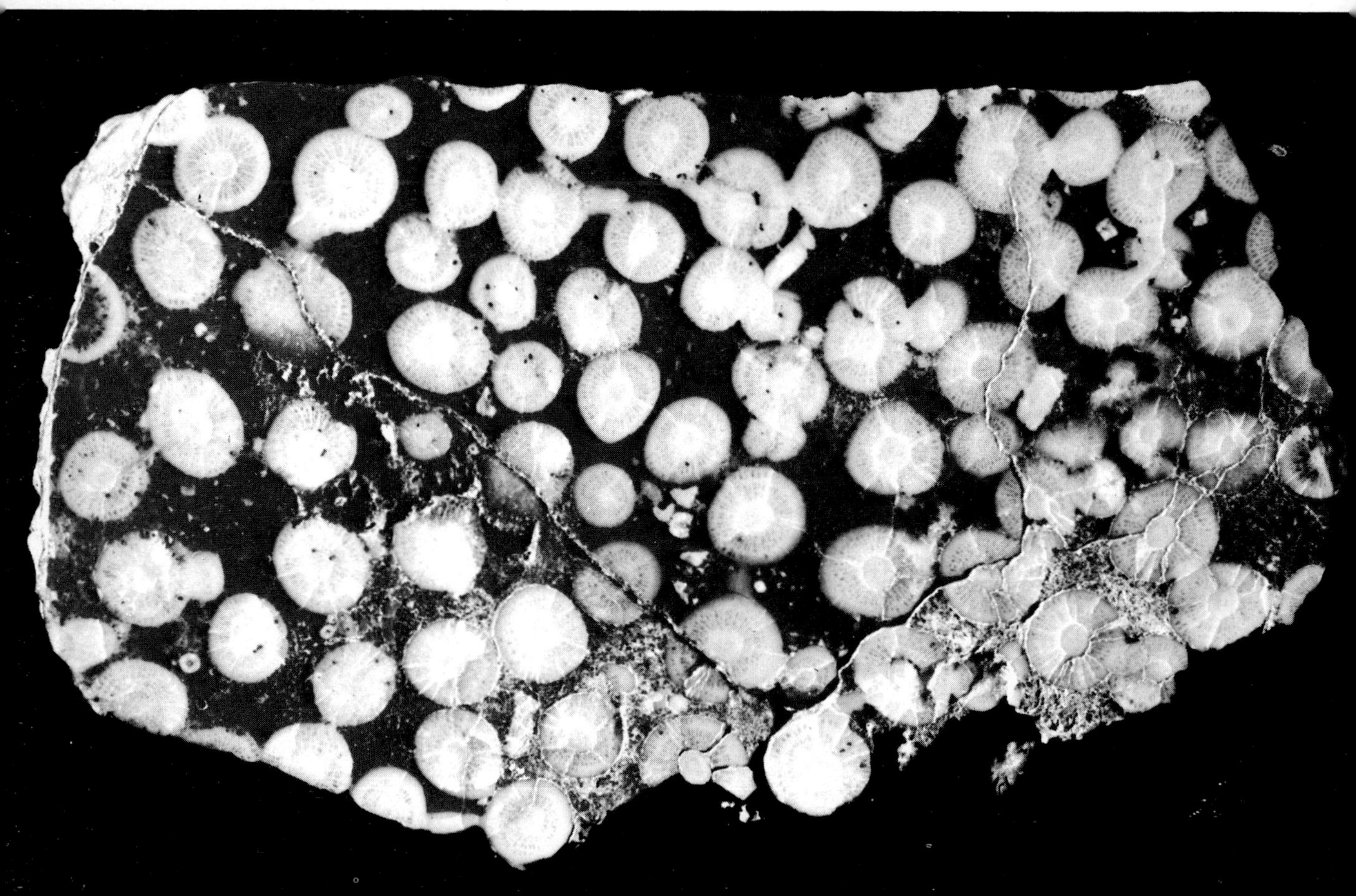

Animals and plants have been used to bolster theories of continental drift, i.e. that the continents originated in one land mass and moved slowly apart. For example, marsupials are found only in Australasia and the Americas; this has led to suggestions of a tie-up of land masses via Antarctica. Marsupials are unique not only in their pouches but in other aspects of their reproductive systems.

TOP LEFT: A Koala bear, one of the most familiar and lovable of Australian mammals. Apart from the mother's pouch, the Koala's cheeks are also pouched to store food.
TOP RIGHT: An Australian cuscus, a distant relative of the North American marsupial, the opossum.
BELOW: Other marsupials: the marsupial mole (RIGHT) and the sugar glider (LEFT).

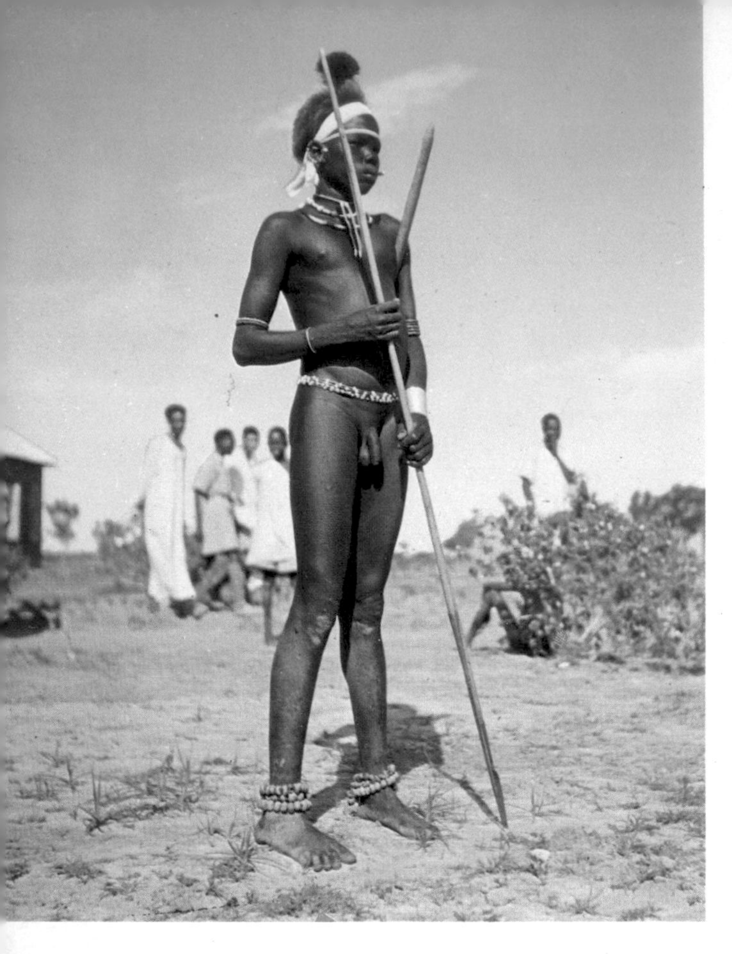

Heat and cold The earth's temperature span is 263°F (134°C), over 80° more than the differential between an ice-cube and boiling water.
ABOVE: Two ways of dealing with the heat: Dinkas in the Sudan (LEFT) strip down; Arabs in Saudi Arabia (RIGHT) follow the example of their sheep and keep the heat out with plenty of insulation.
BELOW: In complete contrast to the Middle East are summer (LEFT) and winter (RIGHT) in Siberia. Even in summer the ground is permanently frozen.
OPPOSITE ABOVE: Tunnelling in the 'Permafrost' in Verkhoyansk, U.S.S.R. Verkhoyansk is remarkable for its contrasts of temperature, ranging from −95°F (−35°C) in winter up to 100°F (43°C) in summer (OPPOSITE BELOW). Hot summers such as these are exceptional for the Arctic circle.

Climate and plants TOP LEFT: The natural treeline occurs close to the 50° isotherm, which links those places whose warmest month averages 50°F (10°C).
BOTTOM LEFT: Plant growth will respond to man-made alterations to the natural climatic conditions: trees growing in an irrigated area in the desert.
ABOVE: Mexican desert: the temperature here can rise to 136°F (58°C) in the shade.
BELOW: This tropical rain-forest need not differ in temperature from the desert above, but the enormous rainfall, rising to over 400 inches a year, makes it a paradise for plants, if not for people.

Local factors can affect climate. An extreme example is shown in this aerial photograph of a village on a cold day, where the frost has highlighted the outline of the old medieval walls. The greater conductivity of the soil beneath the surface where the walls once stood has led to greater heat loss and therefore to greater frost.

Our own solar system illustrates this point. An examination of the various planets can show, in so far as facts are known, the variety of daylength and yearlength existing in this arrangement of nine planets, many moons and one sun. It also shows, again in so far as the facts are known, how many of these orbiting bodies are likely to support life. Their environments may be different, but life is extremely adaptable. For example, it is known that some bacteria and even fungi can live without oxygen, some planets have survived a vacuum by creating their own atmosphere, some bacteria have shown themselves tolerant of powerful acid solutions, while others can live on minerals or survive a pressure of 8,000 atmospheres. Mosses, lichens, and algae have lived for many days in liquid air, −164° F, (−109° C), and their spores have even survived a prolonged dip in liquid helium, −456° F (−271° C). Such extreme conditions, admittedly causing a lack of biological activity in most of these instances, do show the tolerance of living tissue to different conditions and many of them are far more severe than the worst spots on Earth.

Unfortunately, it is all very well, and possibly irrelevant, that life can survive such hardships. It is far harder for life to have originated under similar conditions, and the planets often possess a harsh environment which totally eclipses anything found naturally on Earth. Nevertheless, as our solar system is the only one in the universe about which any facts are known, it is our only planetary yardstick. Its potentiality for life, and the cycles to which it is subject, form the sole guide to conditions elsewhere. Therefore, starting with the outermost, Pluto, the planets can be examined in turn for the relevant facts about their environment, although the most distant of them can be dismissed fairly rapidly.

Pluto's orbit was only calculated in 1915, the planet itself was not discovered until 1930, and its diameter of 3,700 miles was only measured in 1950. Its orbit is extremely elliptical, varying between 2,760 million and 4,570 million miles in each journey around the sun, a journey which lasts for 248 Earth-years. The angle of its axis to its orbital plane is unknown, and so is the speed of each revolution on this axis, but Pluto's colossal distance from the sun (its mean distance away is forty times that of Earth) makes daylength or yearlength irrelevant as it must be far too cold at all times to support life, as we know it.

Neptune, the eighth planet, has its axis tilted at 28° 48′ to its orbital plane (not too different from that of Earth) and it takes 164 years to go round the sun, but it too is a huge distance – 2,793 million miles or thirty times that of Earth – from the radiant energy and warmth of the sun. However, the sunlight reflected to Earth from this place does indicate that there is methane in its atmosphere, and there is an extremely slender possibility that this gas is the product of bacteria living there.

Uranus, the planet discovered accidentally in 1871, also has methane and therefore the bacteria argument can equally well be applied, although the argument is still no stronger. The axis of Uranus's rotation lies at 97° 55′ to the plane of its orbit. Owing to the planet's distance from the sun (1,880 million miles, or twenty times Earth's distance) it takes eighty-four years to complete one orbit. There is that methane, but no one really argues for the existence of life in such a spot.

Saturn's mean distance from Earth is only 883 million miles, and this planet has been studied for a long time. It gave its name to Saturday, and its year is 29·5 Earth-years long. Its axis is at 26° 44′ to its orbital plane, but it revolves extremely rapidly round this axis, so much so that its bulging equatorial diameter is 7,000 miles greater than its polar diameter. No one knows quite how rapidly it revolves, and therefore how long is its day, but that oblateness implies that it must be very short. Like all the planets beyond it, Saturn must also be extremely cold. Its atmosphere appears to be clouds of ammonia plus gaseous methane, unappealing to most forms of life, but, once again, possible products of bacterial existence, however slight the possibility.

Jupiter is Thursday's planet. (In English Thor successfully ousted the Jove of *jeudi* and *giovedi*.) It is 843 million miles from the sun, its axis is only at an angle of 3° 05′ to its orbital plane, it takes 11·86 years to go round the sun, and its own period of revolution – or daylength – is about nine hours fifty-three minutes. Once again such a high speed of rotation has caused considerable polar flattening. The planet weighs more than all the other planets put together, and so has a dense atmosphere. No one knows much about the surface of Jupiter, but its outer atmosphere consists of hydrogen, methane and ammonia. It may not have a surface at all, but only an atmosphere becoming denser and more liquid towards the centre. As for life, despite the presence of these gases, it is as unlikely as on the other four major planets which exist outside its orbit.

Now to Mars, to Tuesday's sphere (as in *mardi, martedi*), to the home of H. G. Wells' Earth invaders, to the red planet, to those 'canals', to our neighbour so frequently considered the most likely home of extra-terrestrial life. Unfortunately, recent work, particularly of fly-by spacecraft, has reduced our estimate of the chance that the planet has life, but it is still the most interesting possibility. It is 142 million miles from the sun, it takes 1·88 years to travel round the sun, the angle of its axis is 23° 59′ (almost identical with Earth's 23° 27′) and its rotation time is twenty-four hours thirty-seven minutes (again strangely similar to Earth's daylength). These various similarities have inevitably made us even more interested. It experiences four positive seasons, although all are longer than ours because of the planet's longer journey round the sun. The lengths of the spring, summer, autumn and winter in the northern hemisphere (prejudice in favour of a northern hemisphere might as well be consistent) are 199, 183, 146 and 159 days respectively. As on Earth the polar caps grow in winter for each hemisphere, and retreat in summer. And, as on Earth, various colorations – greyish-green, brown and reddish patches – also oscillate in tune with the seasons. It is highly tempting for the astrobotanists to think of these patches as areas of vegetation, perhaps like lichens, perhaps something more advanced.

Before the space age the likelihood of life on Mars was looked upon as a probability more than a possibility. Any controversy was mainly over the question of the primitiveness or intelligence of life there. The brown patches reflected light as if they were lichenous areas, and therefore abetted the primitive idea. Conversely the Italian astronomer Schiaparelli, who had reported the existence of 'canals' in 1877, was thereby inadvertently assisting the idea of intelligence on the planet.

Unfortunately not everyone could see the canals, not everyone agreed with the postulation of vegetation, and such dissidents preferred to think of the coloration as mineral in origin. Its waxing and waning with the seasons does not have to involve life because the increased heat of summer might make oxidation of the exposed surface a cyclical event. The planet is farther from the sun than the Earth by forty-nine million miles, and it is certainly cooler. On the Martian equator in the hot season the daytime temperature might reach a balmy 70° F (21° C), but it may well fall to −70° F (−57° C) during the following night. The poles are, of course, much colder. The conditions for life are harsher than on Earth, but they may be commensurate with life. Russian scientists have shown that plants growing at high altitudes on Earth and at low pressures have optical properties similar to those emanating from the dark Martian areas. The astronomer Patrick Moore, attempting to simulate on Earth the conditions then believed to exist on Mars, found that most life forms died in such an environment, but a few bacteria were even able to multiply.

In short, throughout most of this century, Mars has been presumed to possess life of some kind, although simple rather than advanced, primeval rather than Wellsian. The astronomical correspondent of *The Times* of London, having reported in 1955 the appearance of 'bluish-green' colour on what had previously been a Martian desert, then summed up the prevailing views on the planet. Such colourful appearances:

'... were not to be confused with the more regular and predictable seasonal changes, affecting the general intensity of the markings ... The most likely explanation of these markings is that they represent areas covered by some simple form of vegetation, able to exist under low temperatures in an atmosphere of low density. Very little is known of the composition of this atmosphere, but the existence and melting of the two polar snow-caps shows that it must contain at least a small amount of water vapour, such as might suffice to maintain the growth of vegetation of the lichenous type.'

Then came the space age. Then, in particular, came Mariner IV in 1965, and a magnificent flight past Mars. Instead of a distance of some fifty million miles between us and the planet, there was a gap of a few thousand miles as the spacecraft passed by, having travelled for 228 days to get there. The pictures transmitted back to Earth were depressingly good. They caused their distress by standing up on end so many of the previous convictions. For example, the surface pressure on Mars had been generally thought to be equivalent to the Earth's atmospheric pressure at 55,000 ft. Mariner's instruments measured it as 10–20 millibars, or equivalent to a terrestrial 93,000–102,000 ft (over three times higher than Everest, and between a fiftieth and a hundredth of Earth's surface pressure).

The polar ice-caps, which so comfortingly gave the impression of abundant water on Mars, are now less readily assumed to be frozen water. It is thought the ice-caps may be formed of solid carbon dioxide. The amount of this solid gas could oscillate with the seasons just as dramatically as the ebb and flow of solid water. The famous canals do not exist on Mariner's photographs. This was less of a surprise than the revelation that the Martian surface was, in appearance, much like that of the moon, pock-marked with craters, seemingly arid, and much more desolate than had been anticipated. Even the famous 'seas', observed by astronomers for centuries, were far less clear and convincing from close up. No one had believed they contained water, but astronomers had felt they were *something*. Now it appears they are less significant. Even the moon has tall mountains and deep valleys of a sort, but Mariner showed Mars to be without discernible mountain chains or valleys of any note. The flights of two later Mariner spacecraft, VI and VII, had less dramatic results because Mariner IV had stolen the thunder but they both enlarged upon the earlier discoveries. Their pictures made it seem as if the original canals were either ridges or geological fault lines.

The new information is unflattering to previous firmly-held opinions, but it still does not forbid the existence of Martian life. That problem is still extant, and will have to be solved by future spaceships, probably by Voyager in 1971, which is scheduled to make a soft landing there. The thin atmosphere means that any parachute descent is out of the question, and the landing will have to employ retrorockets in the same manner as is necessary for a moon landing. The thin Martian atmosphere also means the lack of a shield against much harmful radiation, and the lack of an appreciable magnetic field means an inability to divert high-velocity charged particles from bombarding the planet. Already some biologists are saying that Mars is, and always has been, and always will be, lifeless. Perhaps so. Or, perhaps, the planet will upset the apple-cart yet again. After all, it would be almost equally fascinating to discover either that life had been there in the past, or was there today, or will be in the future.

Venus is Earth's nearest neighbour on the sun side. It is always covered with some opaque gas, and this has tended to cloud all possible terrestrial convictions. For centuries no one had any idea of its speed of rotation, and therefore how many nights and days it might experience in its 225-day journey round the sun. Radar produced an answer in 1964. Even at the incredible distance of some twenty-six million miles, radio astronomers were able to discover that the Venusian rotation is extremely slow. In fact it is now believed that day succeeds night rather less than twice in each year on the planet. Mariner II, which had flown past the planet the year before, reported that the surface temperature of Venus was about 800° F (430° C). It has been known for a long time that the planet was hot, but nothing like so hot, and so impossible for life, as Mariner asserted. Seasons are therefore irrelevant and, anyway, the axis's angle of inclination is not yet known, although the astronomer Carl Sagan has reported that it is thought to be greater than 7°. So there would be seasons were it not for the fact that such minor climatic differences would be effectively swamped by the intense heat and by a daylength of 120 days. Desert creatures on Earth survive because the cool darkness of night so swiftly follows the heat of the day: a day 120 times longer and considerably hotter prolongs such relief inordinately.

Mercury is the innermost planet. (Vulcan had a good run as a planet orbiting even nearer the sun, but no one believes in Vulcan any more.) Mercury's year is eighty-eight days, and it is believed that the angle of its axis is somewhat greater than 15°. It has a diameter of 3,100 miles, but its mean distance from the sun is only thirty-six million miles. Not only is it rather too close to the sun for organic comfort, but the planet has a very slow rotation. Therefore, as with Venus, each spot has plenty of time to be scorched by the sun before being subjected to an equally lengthy period of cold and darkness. Mercury is undoubtedly too hot for life, and then too cold; it loses on both the swings and the roundabouts. Also it is now thought that the Mercurian day may be two Mercurian years long, an odd state of affairs, but by no means odd enough to suggest that there may be life there. Opinions about planets are tentative things to hold, particularly today, but no one is thought to be chancing his arm should he state positively that Mercury, the final planet, is incapable of harbouring living things.

Similarly, no one is risking his reputation should he affirm, even more categorically, that the Earth's nearest and most beloved neighbour in space has no life on board. The moon must be without it. Its atmosphere is negligible. Its two-week days are seared by the sun – to some 200° F (93° C) – and its equally lengthy nights suffer a temperature drop of 450° F (250° C). The successful landings on the moon and the analysis of the moon samples have already had much to say about the history of the Earth, and therefore about life on Earth, but the moon itself is believed to be sterile. Any organism taken there, which then escaped from the life-support devices fired up from Earth must surely have died. That intensive sterilization programme carried out by the astronauts on their return was partly to allay fears on Earth and partly to act as preparation for eventual returns from the planets, which are far more likely to harbour pathogens than the moon.

Finally, in order to put these Earth-bound assertions and speculations into perspective, and trying to use some of the power satellites have given us to see ourselves as others see us, is it possible to detect life on Earth without actually visiting the place? Assuming that there was such a being, a Martian citizen viewing us from Mars today might be permitted to think that the large amounts of oxygen present in our atmosphere would actually have prevented – by oxidation – the complex molecules necessary for life from arising in the first place. (The answer is that there was not all this oxygen when life began.) The Martian would then have second thoughts and realize that to have both oxygen *and* methane in our atmosphere indicates that both are being produced, however small the quantities of methane. Oxygen can be produced naturally and inorganically, and quite independently of any biological system, but the capable Martian would also calculate that the methane would be able to reduce such a modest production of oxygen. To have an atmosphere of 21% oxygen demands that life is here; the quantity is too much to have arisen inorganically.

Similarly, to the more practical Martian, it must be possible these days to pick up some of our transmissions, and not only the radar waves beamed at Mars itself. Mere vision, either through their telescopes or via the cameras of Martian space vehicles in Earth orbit, might be quite incapable of detecting life. Our great cities, and our greater ravages of the countryside, can be curiously invisible. The Tiros weather satellites have sent back thousands of pictures of an apparently desolate Earth. One exception has been the effects of Canadian loggers who, according to some orbital photographs, cut broad swathes through the trees at right-angles to each other; the resulting criss-cross pattern is both clear and clearly an artefact. In other words, the Martians ought to be able to divine that there is life on Earth but might fail to collect any convincing proof. Should they be able to pick up and evaluate our radio transmissions, they would then learn what we sound like and even see what we look like.

To conclude. The chances are, despite the differing conditions existing on the different types of planets in our solar system, despite the differing diurnal and seasonal periods, despite the variety and forms of life capable of living on Earth, there is probably no other living thing existing near us. Mars was thought likely, but now less so. Therefore the nearest life is unlikely to be nearer than the four light years distance to the second nearest star, assuming, of course, that this star has planets. Our solar system, by having one inhabited planet out of nine, does not necessarily indicate the proportion of inhabited planets in the universe. And, just because a guess has been made that there are at least ten million million planets in the universe, this does not mean that over one million million, or one out of nine, will be inhabited. Why should the pattern here be duplicated there? On the other hand, why should it be unique? The argument in favour of life elsewhere merely takes, as its major premise, that colossal number of assumed planets and the high probability that some proportion of them are inhabited. Even one in a million means the existence of life in ten million different places.

Norman Horowitz, of the California Institute of Technology, has said that 'the discovery of life on another planet would be a monument to our age. Not only would it be an unparalleled technological achievement but it would be a momentous scientific event that would enlarge our view of nature and ourselves and provide unique evidence bearing on the origin of life.' It would be far more than that. It would, in its way, be earth-shaking.

6 Seasonal Life in the Past

The Earth has not always been as it is today. There used to be more days in the year. The temperature used to be, on average, warmer than now, but it has also been colder. The sun has not been regular in the radiation it has sent our way, but increased radiation does not necessarily mean a warmer climate. The actual poles, it is generally assumed, have probably been roughly where they are today, but the magnetic poles have swapped positions time and time again. The mountain ranges have come and gone. Even the continents have shifted, and modern nomenclature is inappropriate for the old land shapes. Warm climate plants lived at one time in Greenland. Coal is found in Spitzbergen, and the prolific growth of vegetation required to make such a coal seam is a feature only of semi-tropical conditions. Conversely, an ice-sheet once covered much of southern Brazil. Even in the past million years of the Pleistocene the northern hemisphere has oscillated from ice age to warm period four times, and it is still warming up. On a different scale, despite this general warning, there have been smaller shifts in the climate: vineyards flourished in medieval southern England, and it then became far colder between 1550 and 1850. Right now, and locally, it may be the peak warm time of one of these minor fluctuations, although a vineyard is still difficult.

So it is not only the days and the seasons which are associated with a changing situation. It is the planet itself. Everything about it is too interdependent for stability to be ensured. Suppose that the sun sweeps up more interstellar gas – which it is constantly doing, but to a varying degree. The more gas it collects the more heat the sun acquires, and the more it radiates. The increased heat arriving on Earth will then, as one of its attributes, cause more moisture in the air. This increased proportion of water vapour will, as one of its properties, cause a greater quantity of cloud. A heavy layer of cloud not only keeps heat in (clear, starry nights are cold) but keeps the sunshine out (clear, sunny days are hot). Whether the cloud is over ice, or warm ocean, or snow, or ordinary land is entirely relevant to the effect of its cover, and the effect will be to make the area either hotter or colder. If the final result is the retreat of an ice-cap this will lead to other consequences, to a rise in water-level, to the inundation of land, to changes in currents. And all these effects will have consequences of their own, most of which will affect the climate and this change will, of course, affect what lives where. Having said all that, it is entirely comprehensible why there is argument over the basic cause of the climatic conditions that were so different in the past. What did cause the Günz glaciation at the beginning of the Pleistocene? How is it that so many tropical animals can be

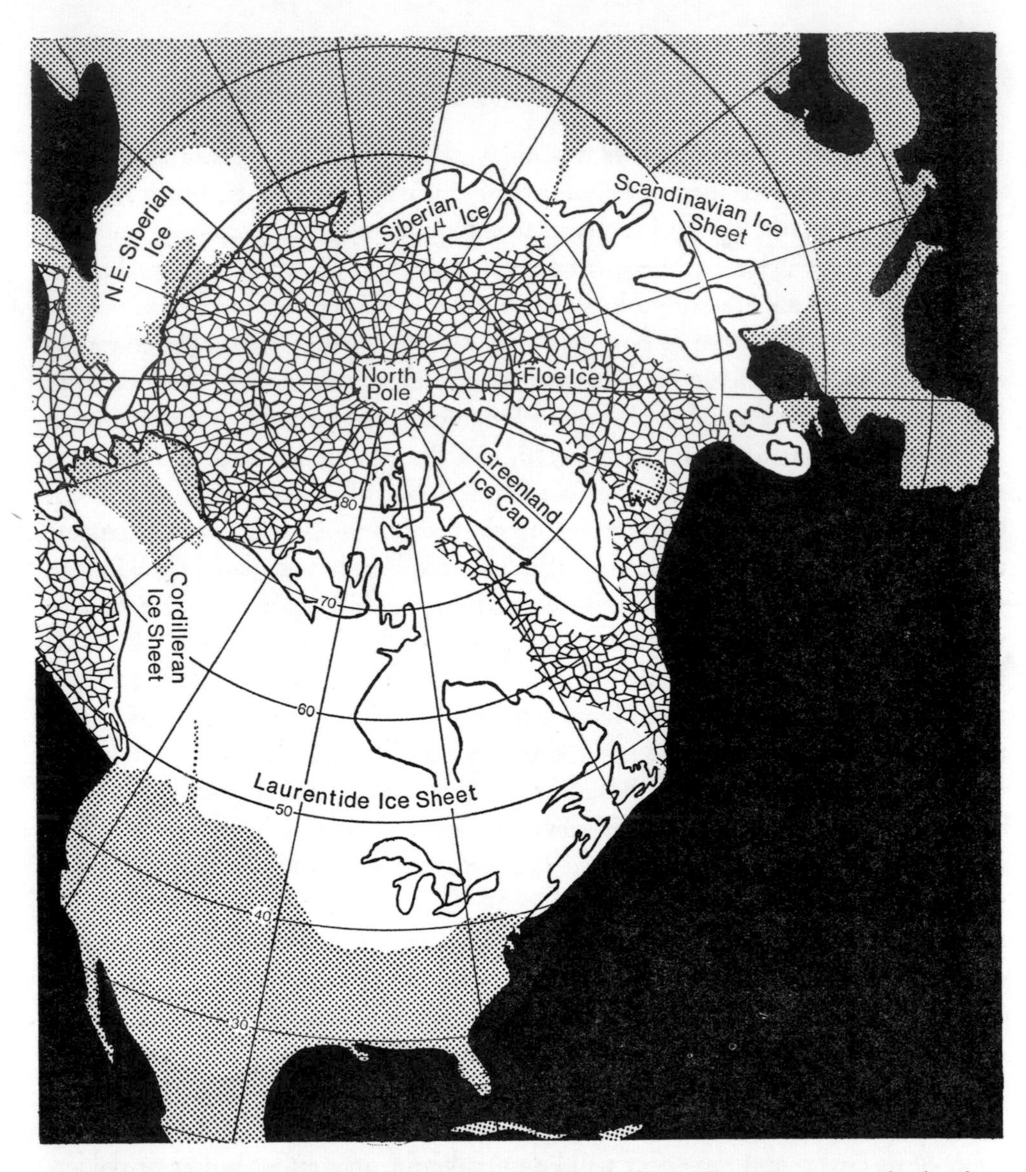

6. Not more than 20,000 years ago, vast ice sheets similar to those still present in Greenland and the Antarctic covered virtually all of Canada and extended as far south as the Ohio and Missouri rivers in the United States, their southern margin lying where summer temperatures are now frequently over 100°F. The polygonal design in the diagram indicates floe ice.

found as fossils beneath central London? What is going to happen in the future, and what did really happen in the past?

The Cambrian age started some 600 million years ago. The Earth itself had been formed 4,000 million years earlier, but by the Cambrian there was invertebrate life, and fossils were being laid down in a far more helpful manner than before. The Earth was then rotating slightly faster than today, and presumably with its axis at a similar angle to its orbit round the sun. The seasons must have been along similar lines, although different in degree. For example, the presence of ice at the poles is generally believed to have been exceptional rather than the rule during and since the Cambrian days. Comparison of the animals living then and now, in so far as this is possible, and with the latitudes in which they lived, helps with the assertion that the Earth was then a warmer place. Today's average global temperature is 58° F (14·4° C), and the prevailing opinion is that our modern times are neither warm nor cold. A frequently quoted average temperature for the bulk of the last 500 million years is 72° F (22° C), with this average falling to 36° F (2° C) during the greatest extent of glaciation. If true, such a temperature range is almost double the difference between the average summer and winter figures in the British Isles.

The arrival of a glacial period must therefore have been a cataclysmic event, as the temperature dropped, as ice sheets advanced, as warm currents were diverted. The English winter is both far less extreme and far less prolonged; yet even it causes considerable change to plant and animal life. An ice age was not just a circumpolar or northern temperate event because temperature dropped everywhere, and the effect on tropical plant growth and cold-blooded life accustomed to perpetual warmth must have been devastating whenever these glacial times occurred. In fact such ice ages have not been common, although the extremely recent retreat of the last glaciation, roughly 10,000 years ago, suggests a greater frequency.

It may or may not be relevant that modern man, *Homo sapiens,* strode on to the scene during the last glaciation, and has flourished and conquered in the time since the ice-sheet retreated to its present position. Should a fifth ice-sheet come – and it is believed there were five glaciations on the previous known occasion – the present so-called Holocene period will turn out to be the fourth interglacial period of the Pleistocene ice age, and just a warm interval before the fifth and final glacial period. It is interesting to wonder whether mankind will then be sufficiently advanced technologically to halt the creeping ice-caps. If not, remembering the fact that much of Scandinavia was beneath 5,000 ft of ice on the last occasion, and that the North American sheet came down deep into the United States, the destruction would be tremendous. If the freezing invasion of ice happens as quickly as its retreat, and we know that to have been extraordinarily swift at the end of the Würm period, even technologically advanced civilizations of this planet will have their work cut out in making the necessary preparations. (Anyone who has not read *The First and Last Men* by Olaf Stapledon, and who is interested in man's long-range future hazards of this kind, certainly ought to do so.)

Seasonally, therefore, the situation has been akin to today's state of affairs throughout the fossil history since the Cambrian. The ice ages have interrupted the fairly smooth motion of events much as lava flows have occasionally laid waste to life near by. The residual areas have always existed as a reservoir so that the wastelands can eventually be restocked. The drama of an ice age is likely to be more devastating than can readily be countered by an organism customarily subjected to lesser extremes of climate: but at least there is some kind of adaptation to a varied existence, and at least there are mechanisms (which include mutations and sexual reproduction) for spreading variability and further adaptation. Without seasonal change all life would have found it harder to survive geological change, the inevitable upheavals and disruptions of an ageing planetary system.

No one knows what effect the magnetic field has had upon climate and life in its entirely contrary behaviour. Fairly soon after the discovery of the magnet it was realized that the magnetic pole wanders, that every year magnetic north is different from true north by a varying angle. Navigators have always had to allow for this difference whenever they have used a compass; but what had never been suspected until 1906 was that magnetic north should sometimes be so off course as to be in the southern hemisphere. In that year Bernard Brunhes, the French physicist, found some recent volcanic rocks magnetized in the opposite direction to the Earth's existing magnetic field. This remarkable discovery, and its confirmation elsewhere, could only lead to one conclusion: the Earth's magnetic field must somehow have reversed.

Further investigation on this subject was even more upsetting. Older rocks than those examined by Brunhes were magnetized in the current direction; still older rocks were again quite the reverse. Finally it was realized that the Earth's magnetic field had reversed itself nine times in the past 3,600,000 years. It seems to have two stable states: one pointing towards north, and one to the south. There is no central position, and no one knows why the switchover should take place, let alone what effect it has on the charged particles surrounding Earth (as in the Van Allen belt) and what effect these have upon our climate. Despite this plethora of ignorance the magnetic swap-over system is fascinating primarily because of the light it shines on yet another great area of darkness, that of continental drift.

To what extent have the continents wandered since time began? Of course there has been mountain building (the Alps are new, the Scottish Highlands are old) since land began, and there have been inundations, and fissures, and lakes becoming seas, but the theory of continental drift rides rough-shod over all of these. It stipulates that today's land masses have been basically the same for a very long time indeed, and that they primarily originated from a central lump of land. Many different conjectures of the ancient situation have been postulated, but there are a few fundamental points which are generally well received. For example, even though the broad Atlantic lies between them, the eastern border of the Americas does appear to match the western shorelines of Africa and Europe. The bulge of Brazil fits into the Gulf of Guinea, the western bulge of Africa fits into the Caribbean, and the complicated outline of Western Europe does not violently disagree with the fairly complex eastern outline of North America. Furthermore, the mid-Atlantic ridge mimics these general shapes and helps to imply that the two land masses bordering the Atlantic ocean have somehow pulled apart from each other.

It was principally a paper of Alfred Wegener published in 1924 (he had first thought of the idea twelve years earlier) that established the notion of continental drift. If the jigsaw pieces of continent are shifted around in place and time they can explain so much. Move them slightly nearer the equator and they are conveniently warmed up, thus explaining how tropical plants could have survived during a particular epoch. Move some regions away from the central land mass later than others, and it is possible to give a reason for both the sameness and the uniqueness of the region's fauna. (Australia is similar in having mammals, but distinct in having no placental mammals, the kind which predominate everywhere else.) Wegener's idea was so good and so handy that occasionally he overplayed its handiness. In 1957 E. J. Öpik, of Armagh Observatory, wrote: 'Alfred Wegener and his followers actually tried to explain in a purely mechanical manner all palaeoclimatic changes' ... and had Greenland 'at present travelling westwards at some 60 ft a year'. It is interesting to wonder from this rebuttal whether a scientist who has 'followers' is more suspect than one who has the traditional 'colleagues', but it is undeniable that Wegener's idea fell out of favour.

Magnetism helped to stand it on its feet again. Studies of the permanent magnetism of ancient rocks do suggest that the continents have moved apart. Madagascar, for example, appears to have shifted to the north-east from its presumed original location tucked into the coast of Mozambique. The fauna suggests that it was connected to Africa in the early Tertiary age, say sixty million years ago, and the distance moved in that time is about 600 miles, representing a speed of one mile every 100,000 years, or six-tenths of an inch a year. Such modest haste is both hard to imagine and to measure, but it was the start of the space age in 1957 which ushered in the first really accurate method of measuring the distances between continents.

Sputnik 1 showed how close the Russians and Americans were in their technology, and also precisely how far apart America was from Europe. The previously chartered distance proved to have been wrong by many yards and it is now accurately known. It needs to be if anyone is to measure continents drifting apart from each other at six-tenths of an inch a year, or more, or even less.

It has always been assumed that daylength must have been shorter in the past. The speed of the Earth's rotation was presumed to be slowing down, and this meant fewer and fewer revolutions, or days, per year. Tidal friction was believed to be the cause, just as tidal friction is believed to have been the cause of the moon's presentation of only one face towards Earth. One body near another causes tides and the friction induced in these tides slows down the spinning. There may have been other factors involved, internal factors perhaps, but tidal friction alone is now known to have changed daylength considerably, even during the fossil age since the Cambrian. Calculations indicate that there were 526 days a year 570 million years ago, and 500 days a year back in the fish era of the Devonian some 200 million years later. It was at first thought that the loss of sixty-three days since the Cambrian was easier to calculate than to demonstrate, but support for this conclusion has recently come from an entirely contrary scientific area.

John W. Wells, of Cornell University, suggested in 1963 that certain fine bands existing in coral were in fact daily growth increments, and there are some twenty to sixty of these bands per millimetre (i.e. each is about a thousandth of an inch). Also it had been known that corals possess far larger bands, each equivalent to a tree ring and showing annual growth. Therefore, Wells decided to count the number of smaller bands in each of the larger sub-divisions. Remarkably his results showed between 385 and 410 days in each Devonian year, as against the 400 days for tidal friction calculations, and an average of 380 days for the Carboniferous period, the geological time which succeeded the Devonian. To help prove the point still further Wells counted the bands of modern coral laid down in modern times. He chose the West Indian *Manicine areolata*, and found an average of 360 growth bands per year.

When two entirely different approaches are made to the same problem, whether by the arithmetic of friction and the microscopic examination of coral or by any other pair of dissimilar factors, any resulting agreement between two such answers is far more satisfactory than two approaches from the same direction. The fact that he failed to see 365 bands on the West Indian coral seems to emphasize, by underestimation, the validity of his technique. At all events the result gives strength to the Devonian figures, and to the disappearance of roughly a month of days in the past 300 million years.

Although the day-loss is sizeable, the enormity of geological time renders it unimpressive. It means one day lost in every ten million years. It might have been thought at first that the loss of days would confound all those forms of life attuned to a year of a particular number of days and, had the loss been sudden, confusion might have been the result. As it is the loss was pathetically slow and there was more than enough time to make the necessary adjustments. One day gone in ten million years means less than a hundredth of a second of change every year.

With 428 days a year back in the Cambrian, and with the Earth taking just as long to travel round the sun – or so it is assumed – the actual number of hours per day was then a little short of 20·5. By the Devonian age daylength had almost reached twenty-two hours. Once again the change looks impressive if the years are forgotten, but there is plenty of occasion for adjustment. One suspects – and more on this subject in the chapter on biological clocks – that all organisms with internal rhythms based on daylength would be able to adapt themselves rapidly even to a grossly changing situation, say with thirty extra days a year. After all, birds have even been known to start roosting when an eclipse has brutally altered the daylight situation by blacking out the sun at midday. Nothing could be more sudden than that.

The actual positioning of the continents in those old days twenty hours long is a matter for great conjecture. Since Wegener published his theory others have attempted their own personal jigsaws. For example, A. L. Du Toit, of South Africa, has tried to arrange things so that the Permian ice-sheet, which existed in the Argentine, southern Brazil, west South America and Western Australia, is more logically placed than on today's maps. So he pushed the Brazilian bulge into the Gulf of Guinea, canted over South America so that Buenos Aires was next to Capetown, shifted Antarctica up into the Indian Ocean, and tilted Australia over so that its northern coast faced west and abutted on to East Antarctica. The ice then appears more reasonable for it occurs only in the lowest latitudes of these continents.

Perhaps Du Toit's idea is not entirely correct; or perhaps the ice was formed in those locations partly because the altitudes were higher instead of the latitudes being more polar; or perhaps the climate was entirely irregular. At all events the general belief today is that the continents are shifting and that they have shifted considerably in the past. The spreading of the sea floor that pushes continents farther apart has been demonstrated. The climate of past geological ages can therefore only be assessed at a fairly crude level. With mobile continents, with upheaval and mountain building, with ice-sheets formerly existing quite near today's equator, and with obvious signs of tropical life all over the place, there was certainly a changing environment, and inevitably a changing seasonal situation to partner it. Those large reptiles were cold-blooded, and they must have lived in a warm climate without very cold winters. They existed for many tens of millions of years, which suggests a warm time of great climatic consistency. They lived much farther north and south than would be possible today. They then died out and their habitats were taken over by the warm-blooded mammals. Was changing climate the crucial factor?

No one knows. What is known is that the long-term stability of the environment can be upset. Nowhere is this more clear than with the last Great Ice Age. It was extremely recent, and it was of considerable extent. Strangely no one in modern times knew or suspected there had been a recent ice age until the beginning

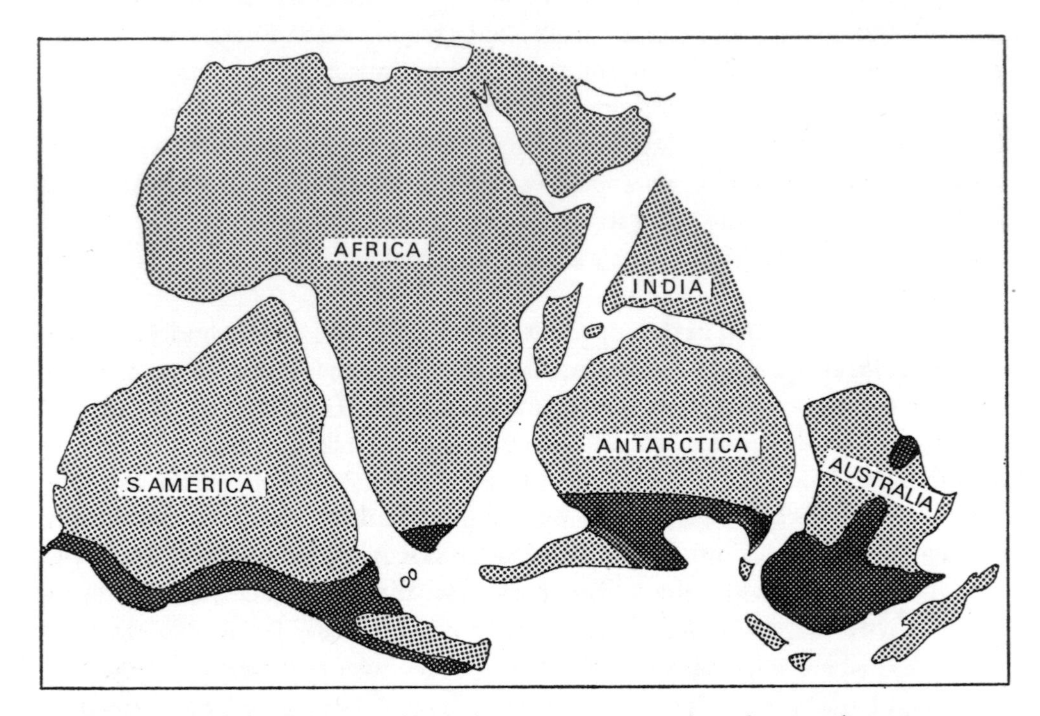

7. The widespread glacial deposits of late Paleozoic time in Australia, Africa, South America and India provided the first great impetus for the theory of continental drift. They suggested that these southern land-masses lay far to the south of their present position at the time of glaciation. Du Toit believed that in Permian times the southern continents were grouped as indicated in this figure.

of the last century. The various signs of earlier ice both in the Swiss foothills and on the continental lowlands had caused people to assume either that the glaciers had been longer, or icebergs had come to rest on the submerged continents or, most bedevilling story of all, that they were further indications of the Diluvium, the grand deluge of the Scriptures. Then in 1829 Ignaz Venetz-Sitten, a Swiss engineer, published his idea of continental glaciation. Louis Agassiz, the Swiss biologist, was critical at first, but then became such a champion of the idea that it became his own. After migrating to the United States in 1840 he continued promoting the idea of an ice age, but never in his lifetime totally destroyed the opposition despite the corroborating American evidence.

However, even he was promoting the idea only of a single glaciation sometime in the million years of the Pleistocene. Not until later, with the initial work being carried out by R. D. Salisbury, an American scientist, was it realized that there had been several Pleistocene ice times with several distinct warm intervals between them. These separate periods were not just the ordinary ebbing and flowing of any climatic or seasonal situation. The intervals between each ice-sheet were long, comparatively warm and entirely clear cut, similar to the period the world is enjoying at present.

Unfortunately, no doubt influenced by Agassiz's switch from Switzerland to the United States, double naming exists for the divisions of the Great Ice Age. Each period is named after the Swiss or American location where the effects of the ice were most clearly seen. The last ice period began with the Günz or Nebraskan ice age which occurred between 275,000 and 800,000 years ago, possibly even 1,000,000 years ago. (Widely differing views are also held about the glacial dates.) The Günz lasted 40,000 years, perhaps more, perhaps less, and was followed by the First, or Early, or Aftonian Interglacial period. Maybe 100,000 years later came the Second, the Mindel, the Kansan Ice Age, somewhat similar to its predecessors. The Second or Yarmouth Interglacial period was probably longer than the first, possibly 200,000 years, and it yielded to the ice invasion of the Riss or Illinois Glacial Period. Another 40,000 frozen years, and then the final or Sangamon Interglacial Age. This interval, probably shorter than the other two, gave way to the Würm or Wisconsin Ice Age, the last – so far.

Right now it is 8–11,000 years since the ice went away. Therefore, even if the current Holocene epoch does lead on to more ice, and if the current time proves to be a typical interglacial interval, the next lot of ice should not creep south for another 90,000 years or so. That respite ought to give technological man sufficient time to prepare whatever is necessary. He should also prepare for a further melting of the existing ice as no one really knows which way the world is going. Nothing is static, but the general warming of the climate suggests that the second of these two periods hanging over us, the melting of ice and the flooding of all ports, is likely to fall first. If the ice age comes first it is not only the cold and the ice that will be a danger; the extra ice-sheet itself will consume so much of the world's water that the surviving ports will then be 100 ft *above* sea-level.

According to some, for instance P. Woldstedt, the ice age itself came at the end of a gradual decline in world temperature throughout the Tertiary. In the Eocene the average temperature was 70° F (21° C). By the Oligocene it was falling, and by the Miocene it was 65° F (18° C). It continued to fall in the Pliocene and had reached 50° F (10° C), when the sudden bouts of glaciation began. Although average world temperature then dropped to 30° F (−1° C) (and remember that this is a postulation, not a fact) it ascended again to 50° F (10° C) within each interglacial period. It was then warm enough (and this is a fact, not a theory) for the African fauna to pour over the land bridges of Gibraltar and Malta/Sicily to repopulate Europe. Hippopotamuses subsequently wallowed in southern England, the virtually hairless *Elephas antiquus* replaced the woolly mammoth, lions killed and hyenas scavenged in the Chilterns and the Pennines. Whereas many species wavered with the ice, retreating and advancing when necessary and when possible, the real casualties of this time were the mammals in general and the largest animals in particular. Almost every major grouping of warm-blooded animals seemed to include a giant form and, whether originally tropical or temperate, these huge rhinoceroses, deer, armadilloes, birds and sloths, for example, are not alive today. Perhaps the advantage of size in a cold climate (greater bulk to surface area) is outweighed by its disadvantages when the warmer weather comes, or perhaps modern man found it easier to kill these enormous forms.

The Holocene epoch started when the ice went, and when the ice-sheets retreated to their present bastions of Greenland and Antarctica. All of mankind's historical time is within this Holocene epoch, and nowhere is history quite so continuously recorded as in China. Detection work among Chinese writings of all kinds (whether this poet called a hill snow-covered, whether that one wrote of pomegranates) has resulted in some tentative conclusions. From 3000 BC to 1000 BC the January temperature averaged 9° F (5° C) higher than now, and it has tended to oscillate from warmer to cooler ever since, with each oscillation lasting 4–800 years. It would seem from this evidence as if there are cycles to climate, for within these major waverings are minor ones of lesser time and lesser temperature change.

Throughout the world the climate is certainly changing, but there is not enough information to say whether it is cyclical. Much of the evidence arises from the marginal areas, and when such a region was sown and harvested. At a symposium held in 1964 on the 'Biological significance of climatic changes in Britain' it was reported that the last 6,000 years have been less stable climatically than the other millennia immediately following the last glaciation. Within this later time temperatures have fallen by 3–5° F (some 2° C), and the treeline has retreated 600–1,000 ft down the hills. Such has been the general trend, but there have been minor variations. It was relatively warm from AD 1000 to AD 1300. (Their vines must have been similar to the vines of today, their techniques not vastly superior, and yet viticulture in Britain nowadays, without modern assistance from glass and heating, is extremely difficult, and impossible in bad years.) Between 1550 and 1850 was the so-called 'Little Ice Age' already mentioned. It is believed that the subsequent warming up started in the eighteenth century, but the cold reservoir of the sea delayed the effects, and the first forty years of the twentieth century were the peak times of this warm cycle. The present time, although past the peak, is benefiting from the same delay as things cool off again.

It might be thought that mankind's current activities, particularly in liberating stored energy, would prevent or at least forestall the cooling; but even the current profligacy in burning reserves of fossil fuel is small by comparison with the 100 million million kilowatts of energy which is Earth's perpetual income from the sun. A greater man-made effect may result from the existence of so much more carbon dioxide in the atmosphere, or of particles of matter, both of which may have considerable consequences. The big man-made geographical disturbances do not seem to make much difference, either for the warmer or the cooler. The colossal Ribinsky dam in Russia, generating heat and creating a large lake, has made a barely perceptible change to the nearby climate. Windspeeds near the lake have doubled, but the effect is only local. Two man-made lakes in the United States, Selton Sea of 400 sq. miles and Lake Mead of 175 sq. miles, both in the dry southwest, have 'resulted in scarcely any change in the climate even in the immediate vicinity', according to C. W. Thornthwaite, of the Laboratory of Climatology in New Jersey.

Towns are customarily warmer than the country, but no one knows whether this extra warmth has any effect on the more general climatic picture. A town is warm not because of all its little fires, or its traffic, or its venous network of a sewage system; the effect comes principally from its buildings, from the way in which their vertical sides receive solar radiation and then retain some of this energy by radiating it across the street to other walls. Cars monitoring the temperature continuously have been driven from one side of London to the other. They have confirmed the temperature rise, often as much as 7° F (4° C), which may mean the difference between frost and no frost, settled snow or none, but this does not mean the climate is being changed, only its local effects. No one can say yet whether the colossal urban areas of the world receive more or less rain, or differing pressures, or any climatic change, as a result of their urbanization. The fact that the Thames used to freeze over, and will not do so under present circumstances, is not related to climate. So many power stations and factories use its flow as cooling water that it is continuously heated above the temperature at which freezing is possible, even in the coldest winter. Besides, the colder the winter, the more power is generated, and the more cooling water is necessary; hence more warming.

A curious frailty of human beings is that each one likes to think his particular day and age is of enormous consequence. Climatic variation in one locality over a few years is assumed to be entirely abnormal, and a portent of greater change. There is also a prevailing tendency to remember one's childhood as an uninterrupted sequence of sunny days, and one's adult holidays as a similar sequence of climatic mishaps. In fact, when the world situation is taken into account, and when large-scale trends are investigated, such local vagaries are swamped, and everything appears more consistent. Globally it would seem as if there has been considerable stability since the end of the last ice age, but there are slow trends which vary for different parts of the world. Within these slow changes are smaller oscillations, which may or may not have a rhythm to them. Emphatically there are not the changes which each man's inexpert eye tends to see in his own short stay on Earth.

However, as some gratification for those who think the current age to be exceptional, it actually is an abnormal time in the whole history of the Earth. Its peculiarities are that the continents are high, that extensive glaciation has only just gone, that the polar world is still ice-bound, that 10% of the land surface is desert, and that it is fairly cool. The Permian period at the end of the Palaeozoic age is thought to have been the last similar occasion. Throughout most of the Earth's history the continents have been generally lower, great inland seas have existed on them, there has been no indication of an ice age, there was probably no permanent ice even at the poles, the climate was certainly warmer and possibly more stable, and there was more similarity to the world's climate as seen from the different regions. According to Prof. Carl O. Dunbar, of Yale, 'the emergence of the continents, the growth of lofty mountain ranges, and the restriction of marine waters appear to have exaggerated climatic zones and regional climatic extremes'. The current world is one of climatic extremes.

So now is an odd time, if the broad sweep of terrestrial history is taken into account, but it is nothing like so odd, or inconsistent, or devastating, or cataclysmic, as many would have us believe. Climate has changed, and is changing, and always will change, both regionally and globally, and no one really knows why. However, it is strange that man's all-pervading and frequently destructive influence on the planet, such as the creation of deserts or his own population growth, has so far failed to disrupt noticeably either the climate itself or the annual modifications which are called the seasons. There is some cause for thankfulness here.

7 The World's Climates

The planet Earth presents an extremely broad range of climatic possibilities. These oscillate with the seasons, but the total range is huge. In Antarctica the Russians have recorded −127° F (−88° C). In San Louis the Mexicans have recorded a shade temperature of 136° F (57° C). Between the two lies a differential of 263° F (145° C), which is 83° F (45° C) more than the difference between an ice cube and boiling water. Rainfall can be twenty-five yards in a year (it being unreasonable to talk of 905 inches), or none at all. Wind can be non-existent for days on end, or a hurricane with gusts of 200 m.p.h. (more about this kind of thing under cataclysms). Moisture in the air – and no living thing is independent of water – can be nil or 100%. Water itself can be fresh, or salty like the sea, or saltier still like the Dead Sea (which still supports life). Cloud-cover can be unknown or perpetual. Ice may be unknown, temporary or permanent. Seasons can be extreme, as in far-eastern Siberia, or minimal. Sunshine can be daily and bright, or six-monthly and relatively gentle.

Greek geographers believed there were seven climates between the equator and the poles. Nowadays it would be ridiculous to talk even of 700 climates; everywhere is a law unto itself, and the law is eternally broken. Most days are either warm for the time of year, or cold. The average temperature for any day is only an average, a rough guide culled from previous years. Everyone living in the British Isles has most positive ideas about summer days and wintry nights, and about the greater cold of Scotland, or the more equable climate of Ireland. Yet, on average, the difference between summer and winter in England and Wales is about 20° F (or 11° C). That pathetic mean difference makes us relish summer and our crops grow, sets us out on holiday and causes birds to lay, it enables insects to flourish and flowers to bloom. To go from an average of 41° F (5° C) (in January) to 61° F (16° C) (in July and August) is from one world to another, a marginal discrepancy of enormous consequence.

Highly relevant to these minute differences between mean maximum and minimum is the growing season. Below 42° F (5·5° C) much of the vegetation of England and Wales just does not grow; above 42° F (5·5° C) it does grow. And with this growth everything else can then flourish. Of course, growth at 43° F (6° C) is still slow, warmer days are best (tropical growth is often so swift it seems almost visible), but the number of days above 42° F (5·5° C) is of extreme importance everywhere. Very cold winters and hot summers are also important, but the length of the growth season – the span of days above 42° F (5·5° C) – is more so. Therefore, spring and autumn can be more crucial than summer or winter. An early spring gives everything a good start, and during the summer it will be growing anyway. A late autumn will postpone the static wretchedness of winter.

In the lowlands of Cumberland the growth span lasts about thirty-seven weeks, with some sheltered spots prolonging it yet further. To go up to Nenthead, for instance, means going up 1,500 ft, and reducing the growth season to twenty-seven weeks. To go up still farther, and to 1,900 ft, which is about as high as anyone cares to have fields in that area, means reducing the season still further. Only from 4 May to 16 October are the days there, on average, above 42° F (5·5° C); therefore growth is twenty-three and a half weeks long, or less than half the year. Not only does this mean that many crops are impossible (all those requiring a longer span) but the grass fails to grow during the colder spell, and substitute food has to be imported for over six months in the year. The main feature of the tropics is not so much that they are hot but that they are never cold.

Seasonal changes still mean a great deal to us even in the towns, although this cannot compare with the time when seasonal change was a matter of growth or nothing, of life or death. No wonder the ancients worshipped the sun, plotted the return of the equinoxes, calculated the date of midsummer, and adjusted their lives accordingly. Today, for the townsman and for the non-agricultural countryman, winter is a nuisance, a time of damp and cold, when some foods may fluctuate in price, and certain food-stuffs will irritatingly disappear from the shops. But there is no longer the total awareness that must have been paramount in former times. Even nowadays in prosperous societies the first day of spring, the first occasion when the sun feels warm on your back, is a time of reverence, of awakening, of splendour. Despite central heating, car heaters, switch-on power, warm clothes, and well-stocked shops, the arrival of spring is still enchanting. Imagine what it felt like for nearly everybody when it meant the return of fresh food and the end of hunger, as it still does today in some parts of the world.

In the developed countries of today, this welcoming of spring is still deeply engrained. Everyone notices the first snowdrop flower, the first cuckoo, the first swallow, and the first lawn mowing. But how many people record, or even look for, the last things of summer with a modicum of interest? Biologically, the swallow's departure is as interesting as its arrival. Does the oak come out before the ash? Often it does, and often it does not; but who cares which sheds the first leaf, and which first bares all its branches to the chill of winter?

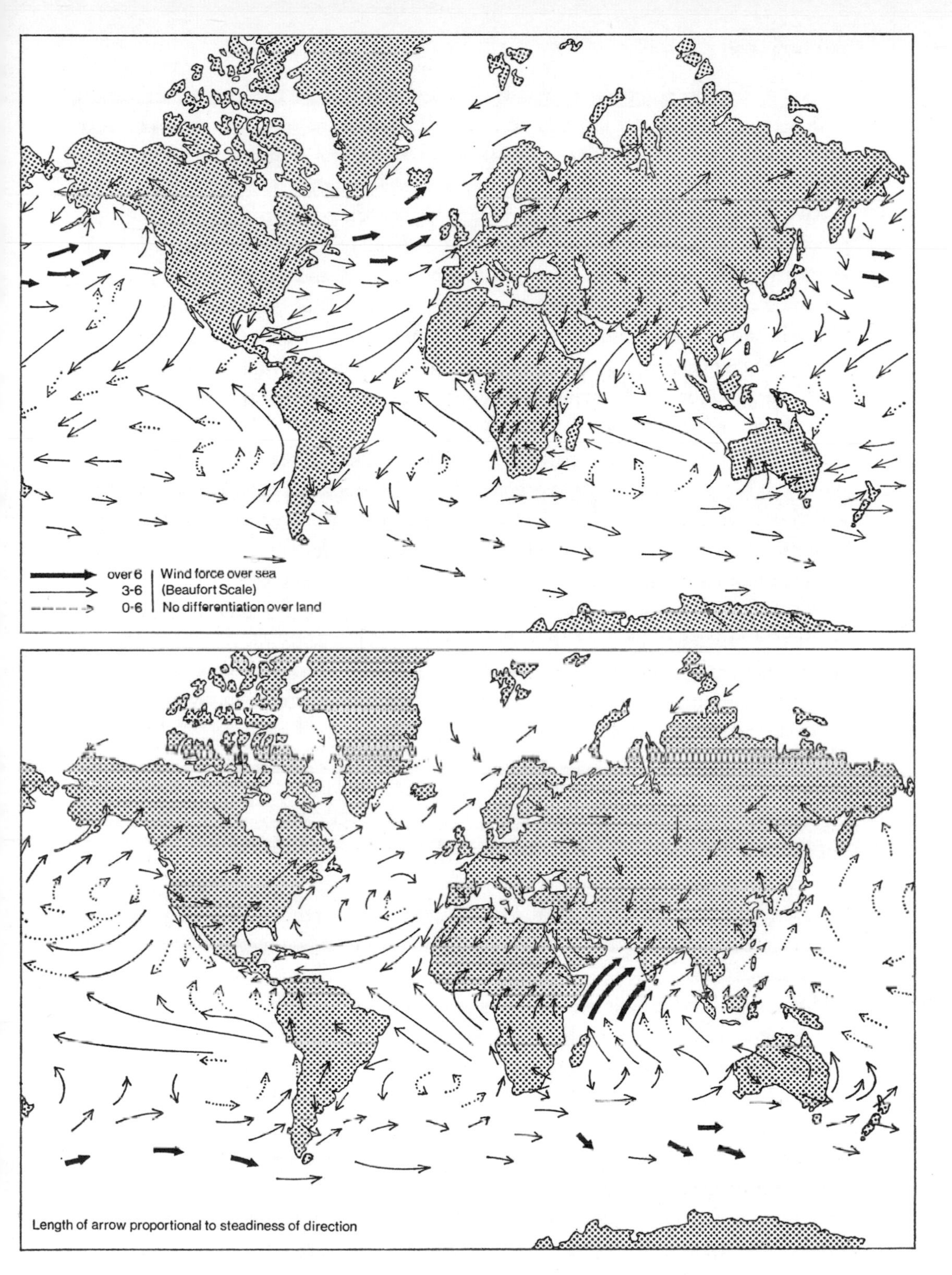

8. In these two diagrams the arrows indicate the influence of land-masses, and the individual pressure systems to which they give rise, upon the theoretical wind 'belts'. The upper diagram describes the January pressures and winds and the lower shows how they have altered by July; only over mid-ocean can the wind directions be easily recognized, while the effects of sea meeting land swing the winds round in a series of swirls around individual highs and lows.

One of the surprising things about the Earth's climate is the amount of variety which exists, despite the fact that the astronomical situation is similar every year, and the angle at which the sun's rays hit any spot on Earth is consistent. Look at any map showing lines of equal temperature, or equal pressure, or equal cloudiness, or equal sunshine, and neither the lines of latitude nor the continents themselves show much relationship with any of them. The system of winds is a good example. Basically, of course, the sun heats up the equator most and the poles least, and the hot air then rises leaving the cooler to take its place. The traditional winds, the westerlies and the trade winds, are the principal streams of air near the surface, but the upper circulation is totally different, and the global picture becomes rapidly confusing. Winds are blowing apparently everywhere, and the broad patterns of direction and strength are seemingly deranged.

However there is a general pattern. Near the ground and between latitudes 30° N and 30° S (i.e. between the two tropics plus a few degrees on either side of them which oscillate between winter and summer) there is a belt of easterly winds, the trade winds. They are less strong both at their boundaries and at the equator, and both boundaries and the equator are good spots for doldrums. On either side of this central belt, and between latitudes 30° N and 60° N (or 30° S and 60° S), there is a similarly broad belt of westerly winds. (Britain, lying within this zone, therefore receives the bulk of its weather from the Atlantic.) Above these surface winds all higher winds tend to become more westerly, and by 40,000 ft the prevailing air stream is westerly almost everywhere, with the band of the equator being an occasional exception. Asia, the biggest land mass of them all, has its own effect upon this generalized system: in winter there is a clockwise flow of air over the continent, and in summer it becomes anticlockwise. Otherwise, and in the northern hemisphere, anticlockwise winds are cyclonic – they always go round a region of low pressure. Clockwise winds are anticyclonic. In the southern hemisphere everything is the other way about. A helpful tag – for some – is that if you stand with your back to the wind in the northern hemisphere the centre of low pressure is on your left.

This general pattern of winds exists despite countless other influences which tend to disrupt it, all the continental situations, the mountain barriers, hot spots, dry spots, wet spots. Everything is interdependent. Hot air picks up more water than cold air. Hot air carrying much water vapour becomes cloudier than dry, hot air when it is cooled. Clouds discourage sunshine from reaching this earth. Denser air may lead to regions of lower pressure. The Earth is spinning and affects the movement of air. So do mountains. So, to a large extent, does land. Air is warmed most at its lower levels. Sea is warmed most at its upper levels. Land warms up quicker than water. Not all land warms up equally, but all land cools off more rapidly than water, and then heats up more quickly the next day; hence the sea breezes and land breezes. Ice and snow reflect heat more than water does. So do deserts more than forests. Rain falls on the windier sides of mountains. And so on. And on.

People often say that a particular day is colder than it ought to be for the time

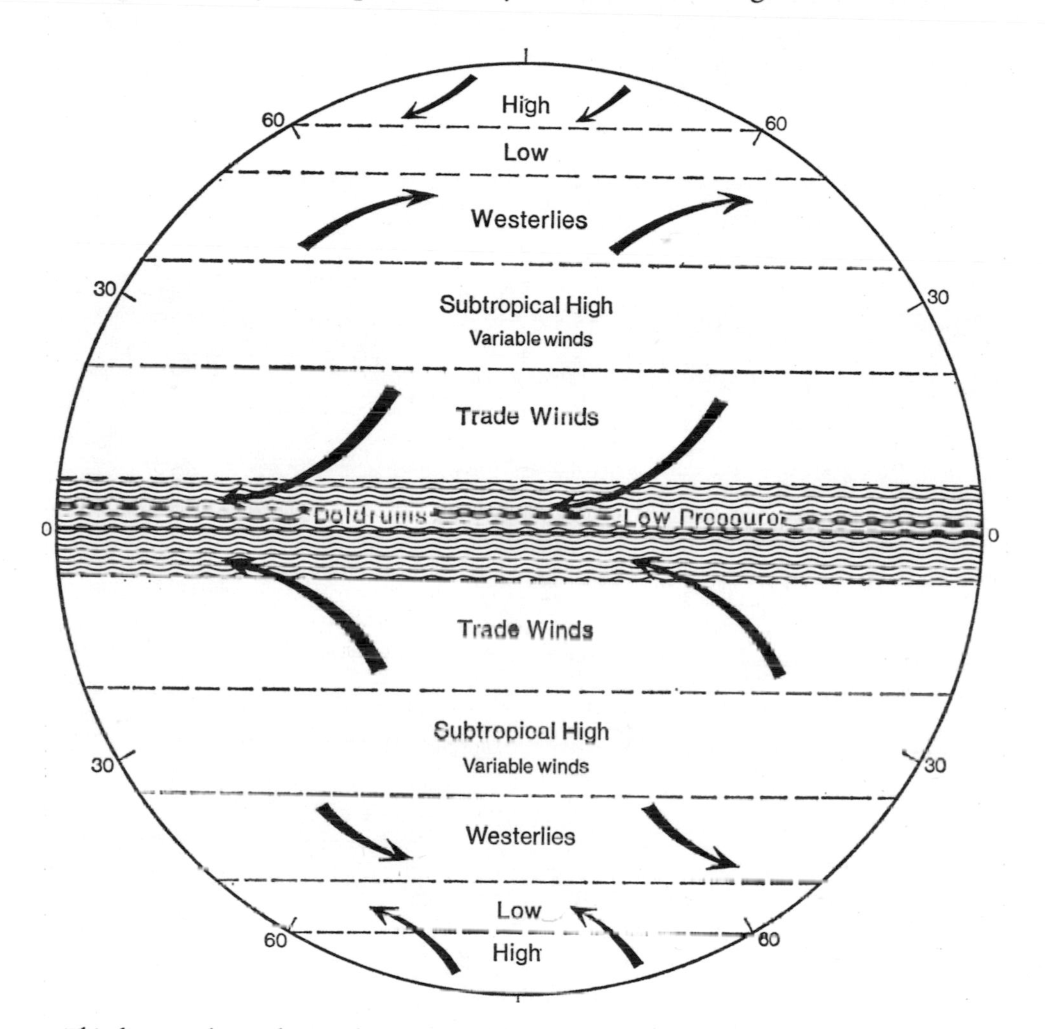

9. This diagram shows the simplest pattern of the pressures and surface winds, such as would be expected to occur on a uniform Earth. Near to the Equator is an area known as the Doldrums, characterized by calm and light variable winds, with occasional squalls, heavy down-pours of rain and thunderstorms. It marks the meeting point of two unusually steady winds, the trade winds, from the south-east and north-east. At about latitude 30° lie the subtropical high-pressure zones which receive generally light winds. North and South of 35° lie the westerlies, which produce changeable weather with abundant depressions and generally temperate conditions. After these follow the sub-polar low pressure belts, and finally the high-pressure polar regions.

of the year. When it is unseasonable they may blame hydrogen bombs just as they once blamed the steam engine for introducing irregularity into the smooth order of things. But to some minds it is more remarkable that any smooth order should

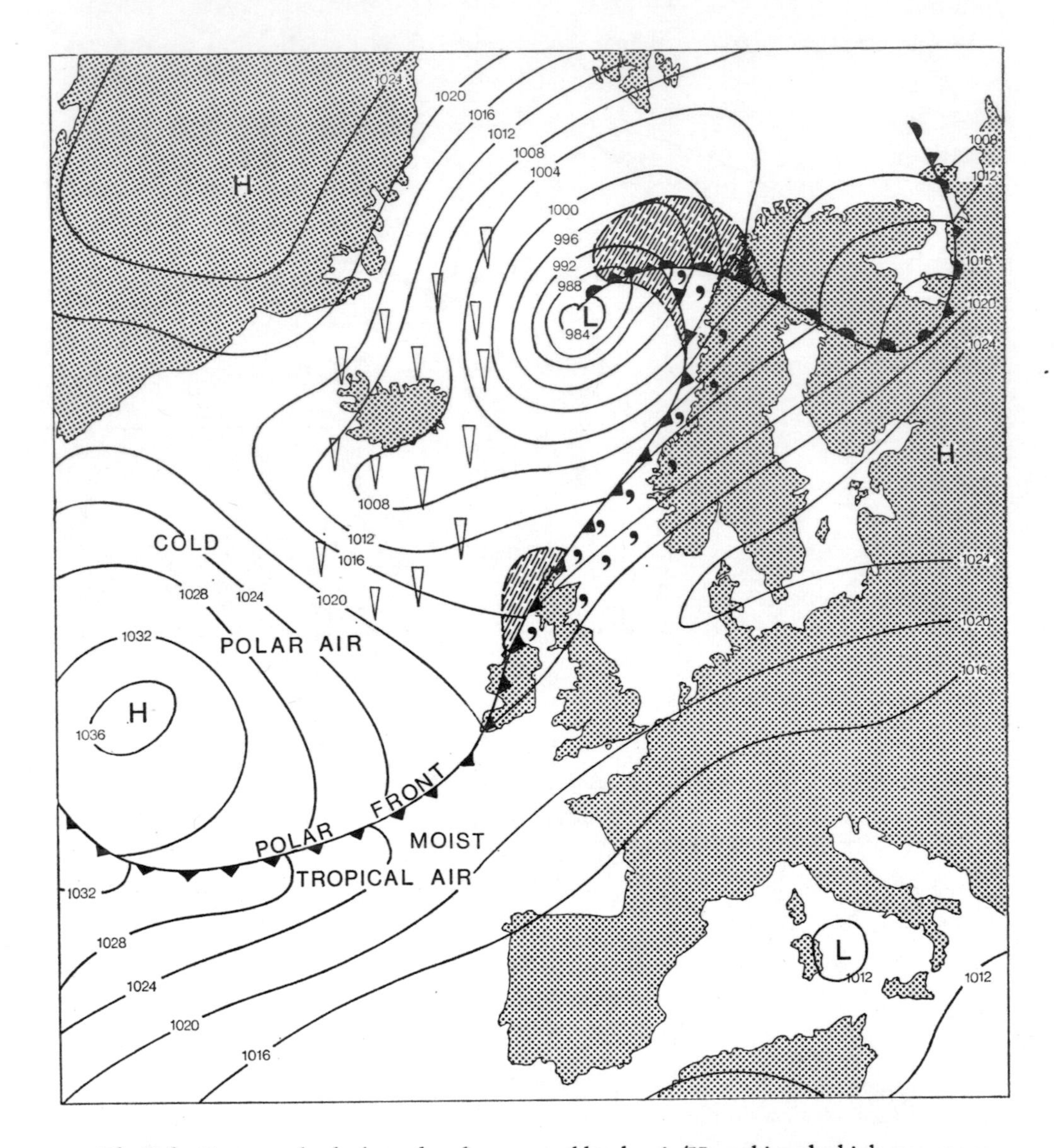

10. The Polar Front marks the boundary between cold polar air (H marking the high pressure or anticyclone round which the winds blow in a clockwise direction) and warm tropical air (L marking the low pressure or cyclone round which the winds blow in an anticlockwise direction). The outbreaks of scattered showers in the cold air, the rather widespread continuous rain near the centre of the cyclone and the occurence of drizzle ahead of the weak cold front are typical features of such a situation.

exist at all, bearing all those interactions in mind, than that an occasional season should be abnormal by a few degrees.

The British Isles are at the receiving end of air streams from Africa (tropical continental; hot summer, warm in winter), from the equatorial Atlantic (tropical maritime; always moist and warm), from Greenland (polar maritime; always cold and dry), from Iceland and the North (Arctic maritime; cold and damp), and from Northern Europe (cold and dry, notably in winter). It can also have mixtures, when two of these air streams are meeting. East Anglia can be the victim of an east wind, and be 20° F (11° C) colder than the west coast of England which happens then to be receiving the westerlies of a tropical air stream. As no air stream travels straight from its place of origin, and as it curves according to the pressure differences, it is possible (apparently anything is possible with British weather) to get Arctic air blowing up the Channel from the south-west.

Not only is the weather interdependent, but there are long-distance links between one period of the year and another. People have their prejudices about these links which may or may not be correct, partly because human beings themselves are poor recording instruments – half an inch of rain spread over twenty-four hours may seem wetter than one inch in a couple of hours. Also human beings cannot tell how moist a week has been (assuming no rain) and, climatically, hot moist days are totally different to hot dry days. Nevertheless, there are some definite associations. Warm sunshine and warm winters in southern England (Kew) are likely to be wet as well. Warm summers are likely to be dry. Cold autumns are more likely to follow cold wet springs.

Unfortunately, such associations are not rules. It would be helpful to planners, to farmers, electricity demand assessors, and the like if there were rules; instead there are only rough guides. It is just that cold autumns are more likely to follow cold wet springs than *either* warm wet springs *or* cold dry springs *or* warm dry springs. In fact the figures are in the proportion of one to two: cold autumns will follow cold wet springs on one occasion for every two occasions that they are followed by the other three possibilities.

If this explanation is indecipherable it is perhaps better, as an alternative, to think of prevailing winds. A wind is said to be prevailing if it blows more frequently from one direction than any other single direction. If it blows for two-fifths of the year from the south, one-fifth from the west, one-fifth from the north and one-fifth from the east, the prevailing wind is said to be from the south. On two days out of five it will blow from that direction, although on the other three days it will blow from one of the other three directions. The same is true, but marginally more complex, for the prevailing association between cold autumns and their cold wet springs.

In a sense climatologists all have slightly split minds. They spend their time trying to discover the broad picture and then, having done so, spend time evaluating all the inaccuracies inherent in the broad picture. The broad wind pattern undoubtedly exists, but it is only barely credible if a world map is covered with all the winds for any one day. Similarly it is well known that the tropics are hot, the temperate zones less so, and the polar circles least of all. But Verkhoyansk, lying within the Arctic circle and on a more northerly latitude than northern Iceland, can reach 100° F (38° C) on a hot summer's day. (British weather hits the headlines, roads become sticky and everyone gasps with amazement whenever the thermometer goes above 80° F (27° C).) At Verkhoyansk a cold night in midwinter has been known to reach −95° F (−71° C), far colder than the north pole itself which lies 22° farther north. (British winters hit the headlines, rail and roads become chaotic, and everyone shivers with amazement, when there are a few degrees of frost and it is still 120° F (65° C) warmer than Verkhoyansk.) In fact Verkhoyansk's ability to range through 195° F (some 110° C) (about three times Britain's extreme range) happily confounds broad pictures of the Earth's climate.

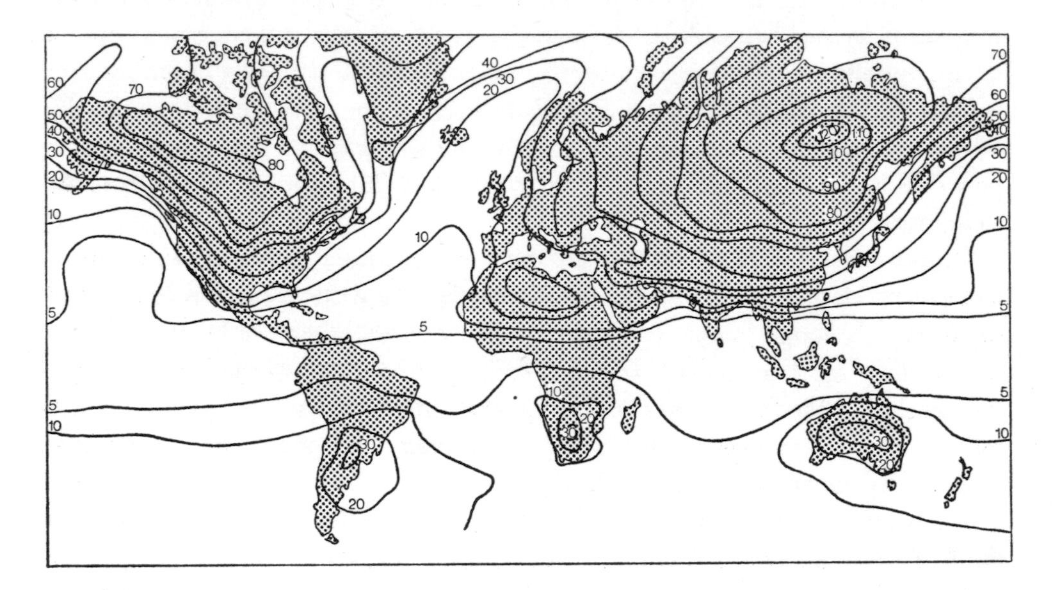

11. This diagram of the mean annual range of temperature shows the contrasts between the two hemispheres. Nowhere in the southern hemisphere does the range exceed 40° whereas in the northern hemisphere huge areas like Canada experience more than double this temperature range and Siberia more than treble.

The Earth's rainfall has similarly bizarre features. In general, and according to the broad outlook, there is most precipitation (snow, hail, dew, hoar-frost and rime all have to be included, and rainfall implies only one kind of precipitation) along the equatorial belt. The north of South America has a lot, so does the Gulf of Guinea, the Congo and the Malay archipelago. The poles have very little; north of 70° N and south of 70° S precipitation is (generally) less than ten inches a year. Globally, there is also a seasonal oscillation in that the heaviest rain in January is south of the equator, while in July it is to the north of it.

However that is only the broad picture; exceptions abound. The coastal area to the west of northern Chile, Peru and Ecuador receives next to no rain; farther south, and to the west of southern Chile, the rainfall can be 100 inches a year. All of the Sahara, the Arabian peninsula and southern Persia receive extremely modest amounts of rain; farther east, but on precisely the same latitude, and therefore equally far from the equator, lies Assam with its tremendous monsoon rainfall. Cherrapunji, in Assam, is often credited with the world's record annual rainfall. The average there is over 400 inches, and the bulk of this falls in the six months from April to September. (One annual total was 905 inches or 75½ feet of water. Probably 90% of this fell in the rainy season, which means thirty inches a week for six months on end, or Manchester's annual rainfall every week for half the year.) Conversely – and such perverse comparisons make nonsense out of average figures – there is that area in the Chilean desert which has not received any rain for over 400 years. Both spots, the wettest and the driest, are about equidistant from the equator.

Cloud-cover, a vital feature of climate, pays even less regard to the world's lines of latitude. Everyone knows that Britain is a cloudy place but its mean annual cloudiness is less than seven-tenths of the sky covered. Cloudier areas are the North Atlantic, the mouth of the Congo, southern Alaska and west of Peru. The cloud pattern is enormously influenced by the presence of the continents, but there is a basic association between cloudiness and the low pressure areas both on the equator, in the temperate areas and at the poles.

Sunshine might be thought of as the one exception, the one climatic factor which bears a time relation to latitude. After all, the Earth's axis is constant, the Earth's orbit is constant and the sun's rays do indeed approach each part of the Earth at a given angle and with a given intensity (solar perturbations permitting) at any particular time of the year. This seasonal variation is a steady variable. However, the amount of sunshine, and the intensity of that sunshine reaching any spot on Earth presents a haphazard picture much like that either of rainfall, cloudiness or temperature. Within the tropical zone the maximum recorded sunshine falls, not on the equator as might be supposed, but on the two tropics – on the tropic of Cancer and the tropic of Capricorn. A midsummer day on each tropic is thirteen and a half hours long and the sun is directly overhead – an impossible combination at the equator. Of course at midwinter there is less, for then the tropic day is only ten and a half hours long and the sun reaches a maximum altitude in the sky of only 43° – another impossible event for the equator.

This dual relationship involved in sunshine, the length of day plus the angle of the sun, makes the degree of sunshine at any one place less simple than might be imagined. Seasonably it is relatively straightforward. The equator receives its minimum quantities at midwinter and midsummer, and its maximum amounts at the two equinoxes in March and September. The middle latitudes receive least in winter and most in summer. The poles receive none in their winters, some during the summer half of the year and most at midsummer.

So much for simplicity; now for the full relationship between the angle of sun and the length of day. Assume it is midsummer's day in the northern hemisphere. On the equator there is then less sunshine than at any other time excluding midwinter. At the tropic itself (23½° N) there is, as already described, the powerful combination of a vertical sun at midday and a daylength of thirteen and a half hours. To begin with, when proceeding farther northwards, the steadily declining angle of the sun (causing the sunshine to be less powerful) is more than compensated for by the increasing length of day (causing sunshine for a longer period). A maximum is reached at about 43½° N, the approximate latitude of Toulouse, Marseilles, Florence, Toronto, Milwaukee and southern Oregon. At this latitude in midsummer the sun is still reasonably high in the sky, and the days are fairly long.

North of 43½° N, although daylength is longer, it cannot compensate sufficiently for the decreasing height of the midday sun. Therefore, with still lengthening days, but with a falling sun, the amount of sunshine continues to fall until latitude 62° N is reached. At this point, the approximate latitude of southern Iceland and Greenland, and 2° N of Bergen, Oslo and Helsinki, there is a minimum quantity of sunshine (still remembering that the date is midsummer's day in the northern hemisphere). Oddly, whatever one's previous convictions, there is more sunshine north of that latitude of 62° N. The daylength now gets so much longer, reaching twenty-four hours near the pole, that it more than compensates for the declining midday height of the sun. A final fact, which causes the midsummer polar sunshine to be so strong, is that even the midnight sun is 23° above the horizon.

Thus insolation, as sunshine is known to the climatologists, is far less straightforward than might be imagined. Even without the mention of a cloud, and sticking solely to one day – that of midsummer – the mutual powers of daylength and sun-height wax and wane most complicatedly. Daylength, of course, influences the daily total of insolation and, equally naturally, the sun is weaker when it is at a lower altitude, partly because its angle from the ground causes the sun's rays to pass through a greater quantity of atmosphere. This absorption of energy is greatest at the poles, where the sun never gets high in the sky, and it has been estimated that only 18% of the sun's energy aimed, so to speak, at the two poles actually reaches them. Elsewhere the proportion is higher.

The effect of water-vapour in the air, particularly when it manifests itself as cloud – and by no means are all types of cloud equal in this respect – has already been described. Quite apart from cloud there is dust. This cuts down insolation (though it is good for colourful skies and sunsets). There is also the question of terrestrial altitude, which means less atmosphere for the solar rays to pass through before reaching the ground. And then there is the nature of the ground itself, whether it is ice (which instantly reflects a lot of the energy) or snow (which reflects more) or sand (which absorbs a lot) or dark soil (which absorbs far more).

The sun's rays are undoubtedly the core of any appreciation of the Earth's climate; but, although they arrive in parallel fashion, their extraneous power is torn this way and that by terrestrial influences. Without an atmosphere, or dust, or water-vapour, or mountains, or oceans, or differing forms of surface, the story of insolation would be reasonably direct. As it is, the sun's constant precision in aiming radiation at us is constantly upset the moment its energy approaches Earth. Hence disruption, hence Verkhoyansk, hence Cherrapunji.

Finally, and as a further debunking of any broad climatological pattern, there is the crucial extra of microclimatology. This can be introduced by describing the Stevenson screen. When the meteorologists of England were beginning to unite, they wished to unify their recording techniques. Everyone knew that thermometers placed in the shade, or in the sun, or near the ground or far from it, would give different readings, and everyone – boosted by the infant Royal Meteorological Society (founded in 1850) – wanted conformity. Thomas Stevenson, lighthouse engineer and father of Robert Louis Stevenson, designed a box known today as the Stevenson screen. Essentially it keeps out all direct sunlight and much reflected sunlight, while its twice-louvred sides permit full passage of air. It also keeps out rain and snow. A double roof painted white helps to keep the air within this box from being overheated by the box itself, and the box is a certain distance above the ground with the ground itself preferably covered by grass. All this is in an effort to reduce local influences, to acquire a true picture of air temperature, pressure and moisture content. The meteorologists who started to adopt the screen after 1866 had known since they were small boys that there is all the difference in the world between running with bare feet on a bright sunny day over grass, asphalt, dry sand, wet sand or rocks.

And, of course, in one way or another all living things know this too.

Microclimatology is the study of the extremely local and small-scale changes in the climate. It is relevant to a human being whether a wall is keeping the wind off, whether a tree is providing shade. It is of enormous consequence to living things whether their niche in life possesses the right climate. A lizard can bask contentedly on a granite slab, drinking in the heat of a burning summer's day. A soft and vulnerable invertebrate would die on that rock, but can hide under a couple of nearby leaves which provide entirely suitable conditions. An animal can burrow a couple of feet into the soil and scarcely know whether it is summer or winter, let alone night or day. Climate is what (with a little ingenuity) you make it, in so far as this is possible.

Man is the most ingenious animal of them all. He has not been able to disregard bad weather, but has adapted himself to the differing conditions of most parts of the globe. (Antarctica was the big exception until recent times.) It has been said that mankind is a tropical animal who carries his own warm climate with him. Only in the hottest places does he not wear any warm clothing and, even then, he is probably under some form of cover at night. Strangely, although clothes of a kind are the rule in temperate and polar regions, there are two totally different approaches within the tropics. One is to wear nothing, like the Dinkas of the Sudan; the other is to wear a lot, like the Arab and the Saharan desert people. (Ordinary temperate humans also divide into two camps on a hot day; those who think it best to strip as much as possible, and those who like at least some clothes to protect them from the fierce blast of the sun.)

Studies of Merino sheep have helped to explain the Arab approach more than that of the Dinkas. It is traditional and wrong to sympathize with an unshorn sheep on a hot day. If that sheep is shorn its breathing rate is subsequently doubled showing that greater cooling is more than necessary. Experiments were carried out at Julia Creek in the middle of the hottest part of northern Queensland. In still air at midday the unshorn, woolly, Merino backs were almost too hot to touch. The difference in temperature between the scorching outside of the wool and its cool depths was 60–70° F (33–39° C). Actual skin temperature was 106°F to 108° F (41–42° C), still hot but far cooler than would be the case without the wool. These sheep can therefore be described as temperate animals which carry their own temperateness around with them. It would be nice to know the skin temperatures, on a hot day, of both Dinkas and Arabs, particularly in view of the fact that white is a better reflector than black, and the Arab burnous is often white whereas the African skin is always black.

Apart from an animal's own portable climate there are all the more static differences near the ground. Dr C. H. B. Priestley of Australia, has pointed out that to go from one centimetre above the ground to one and a half metres above it is often equivalent in its temperature range to a journey from sub-tropical to

The polar regions The polar bear is one of the animals most obviously suited to Arctic conditions, where a mammal needs ten times more heat, or insulation, than in the tropics. It even has hair on the soles of its feet to help it grip the ice more securely. The Arctic, unlike the Antarctic, supports plenty of land-based animals, including foxes, hares, caribou and wolves.

LEFT: Man is little suited to polar survival, and faces such hazards as frozen corneas.
ABOVE: A Ross's gull, a bird which has the perverse habit of migrating in search of colder weather. It nests by the River Kolima in Siberia and then flies north for the winter to Spitzbergen and the polar basin.
BELOW: Penguins are found only in the southern hemisphere. In fact, only two of the world's eighteen species (including the Adélie penguins shown here) breed in the Antarctic.

TOP LEFT: The husky is one of the few animals capable of adaption to the rigours of an Arctic winter.
BOTTOM LEFT: Snow buntings, the last land-based birds to leave the Arctic to fly south for the winter.
TOP RIGHT: Eskimo children in North West Territory, Canada, standing on their igloo roof; the igloo is stronger and warmer than a tent and can be put up in under three hours.
BOTTOM RIGHT: Seals come ashore for the breeding season, but spend most of the year in the sea; their flippers, unsuited to land, make progress tortuous.

After the long winter: the sun returns, rivers melt, plant and animal life begin again, and a year's work is compressed into a few months of feverish activity. But although the days are warmer, at night the temperature never rises above freezing point.

LEFT: The spring thaw of the River Kolima in an unfriendly but still inhabited area of Siberia.

BELOW: Eskimoes in their summer clothes.

RIGHT: Drying meat outside a tent. The caribou skin tent replaces the igloo for the summer. The sun-dried meat is stored for the winter and, provided it stays dry, can be consumed a year or two later.

FAR RIGHT: Duck-hunting. When the rivers thaw, ducks take the place of seals as Eskimo prey.

BOTTOM RIGHT: Eskimo diet consists mainly of fish and meat because of the scarcity of edible plants. 'Eskimo' is said to mean 'eater of raw meat', a term apparently applied in scorn by the Algonquin Indians.

The sudden emergence of the Arctic flowers—in May in the south and in June in the north—is quite unlike anything in the temperate world. Insects breed and multiply, most birds' eggs are hatched by the end of June and the whole natural food-chain springs into action.

TOP LEFT: Diapensia, one of the spring flowers which grow in Interior Alaska.
BELOW: Another Alaskan spring blossom, Cushion Pink.
TOP RIGHT: Snow buntings back in the Arctic after their winter migration. They reach the most northerly lands by April and are normally at their nesting sites by mid-June.

sub-polar regions. Similarly, and when the sun has set, the temperature range can be just as extreme, but reversed. Therefore an insect can experience a whole year of climate in a single day.

Every farmer has appreciated microclimates, that this field is only good in patches, where it is sheltered from wind, where it fails to catch the frost, where it sees the sun. In the past all houses were built with these factors very much in mind. Only in modern times is it possible for a house-builder to start work with scarcely a thought for any of them, leaving the future occupants to discover that they exist in a wind trap of formidable unpleasantness. W. H. Hogg of the Meteorological Office has worked on frost donor sites and frost recipients. A particular frost pocket may, if the slope of the ground is more than two degrees, get rid of all its frosty air down to the less fortunate frost recipient. In between lies a frost path and, possibly, a brand new house. House-builders may consider the sunny southern side of the house to be warmer, for this is generally the case, but it may also be unwittingly in the path of that torrent of frosty air. Or, even worse, it may be in the actual receiving zone of all that air, good for meat storage, bad for human happiness. Certainly anyone planting frost-sensitive crops, such as fruit trees, should be wholly aware of local conditions, and how they can be manipulated to advantage.

The differences can be acute and just as marked on a smaller scale. An aerial photograph of a field on a cold day can show up the former outlines of a medieval village because frost has settled dominantly above the old walls. The soil there may appear entirely normal on the surface, but the better conductivity of the old wall remains beneath the surface leads to greater heat loss and therefore to greater frost. The effect can be most striking.

A good gardener also knows that loose dry soil warms up well during the day but loses its acquired heat more rapidly at night. Firm, compact and undug soil warms up less by day, but keeps warmer by night. Seeds germinate well in the warmth of the open soil, but seedlings need the night-time warmth of firm soil. This is in fact precisely the same variant that exists between the continent and the sea. The land masses, like open soil, warm up quickly, and cool off quickly; the sea, like compact or very wet soil, takes a long time both to heat up and cool off. Similarly, although not for reasons of conduction, cloud-cover has the same effect. It stops the day being so hot, and stops the subsequent night being so cold.

On any one day, therefore, and in any one area, there are a host of different climates. A stand of grass is much like a sheep's back: it is cool and moist in the depths, hot and dry on the surface. On a typical moist, sunny English day in summer an insect in a meadow has a variety of different conditions to select from, and all can be achieved by climbing one stalk of grass.

Assume the grass to be about two feet high, and the insect to have preferences concerning windspeed, temperature and water-vapour pressure. Windspeed, assuming a gentle flow over the land as a whole, rises rapidly with grass altitude: at four ins high it is virtually nil, at eight ins it is about 10 ins per second, at twelve ins it is 24 ins per second, at sixteen ins it is 44 ins per second. Temperature first rises, then falls off: in the depths it is 63° F (17° C), at four ins it is 66° F (19° C), at eight ins it is 70° F (21° C), at twelve ins it is 71° F (22° C), but at two ft it has dropped to 69° F (20° C). Water-vapour on this typical day starts off at 18 millibars, falls to 17 mb. by twelve ins, to 15 mb. by sixteen ins, and is only marginally lower by the grass's limit at two feet. Therefore, from windlessness to 2½ m.p.h. from 63° F to 71° F (17–22° C), from 18 to 15 mb. of vapour pressure, the insect can choose the conditions most suitable for its existence. Human beings, or cows, will tramp through that grass, oblivious of the variegated microclimates at their feet, but both of them can be wholly intent upon discovering a warm, sheltered spot for themselves, a form of microclimate more suited to their larger frames.

Bearing in mind the sensitivity of vegetation to unwelcome conditions, however unseasonal and abnormal, and remembering the greater survival of particular species in particular climates, it might be assumed that vegetation provides a better and more permanent recording of recent climate than meteorologists and climatologists could ever provide, however densely and accurately they might plant their Stevenson screens.

The tree-line, for example, the most northerly limit of trees in the northern hemisphere, might be thought of as a fair guide to the prevailing situation. South of it the trees thrive; therefore warmth must be sufficient. North of it they do not exist; therefore cold must be excessive. This tree-line is similar to the 50° F (10° C) isotherm, the temperature line joining all points whose warmest month averages 50° F (10° C), but this isotherm is sometimes north of the tree-line and sometimes south of it. It appears to be more important to a tree that three months are above 43° F (6° C) than that one month is above 50° F (10° C). And of course it is important to trees whether the temperature is just above 43° F (6° C), or higher. Therefore if the month/degrees are calculated, if the average number of degrees Fahrenheit higher than 43 (6° C) are added up for each month, the result is of more interest, both to scientists and the trees. Assume the three warm months of summer have averages of 44° F (6° C), 50° F (10° C) and 53° F (12° C), the month/degree factor is then 1+7+10=18 (0+4+6=10). This particular factor of eighteen month/degrees, almost irrespective of the manner in which its total is reached (perhaps 6+6+6 (3+4+3), or even 3+13+3 (1+8+1), but not 0+18+0 (0+10 +0)), does tally fairly well with the tree-line. The association tends to go awry near an ocean and during peak years; but, taking into account the perpetual

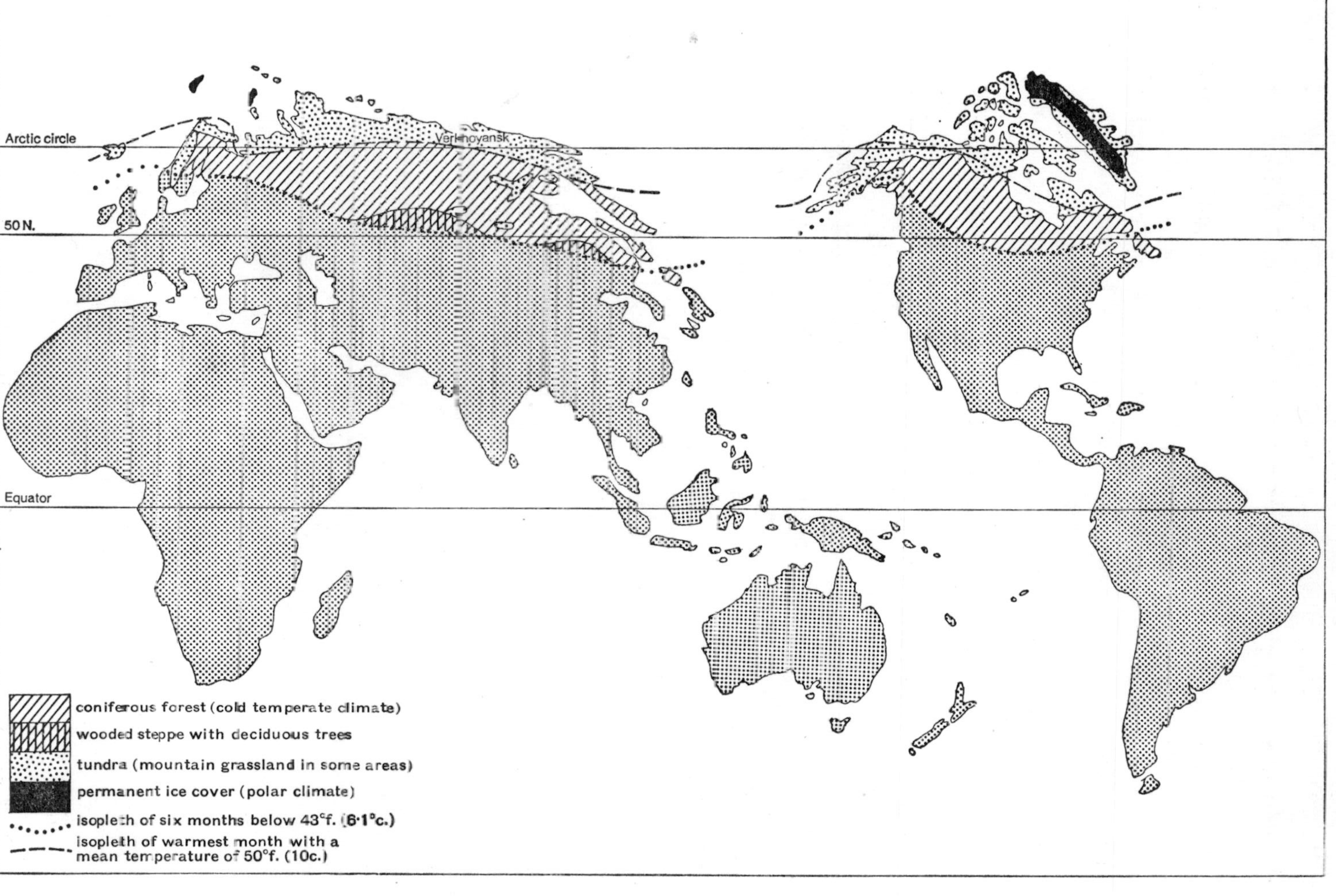

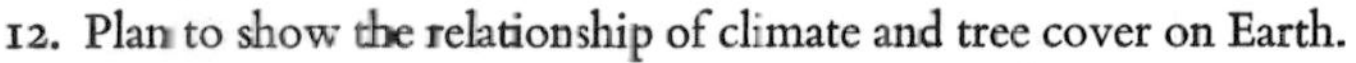

12. Plan to show the relationship of climate and tree cover on Earth.

difficulty of partnering one cause with one effect in the world of climatology, the relationship is outstandingly close.

In a sense it is always unfair to describe freak years, and it might be more reasonable to talk only of normal years, but freak years with their unseasonable cold spells and frosts can be vitally important to the plants. It does not concern them that normal conditions are generally higher; the abnormal frost can still darken their extremities (as strawberry and potato growers know so well) and spoil the crop. The freak occasions are important but the more prolonged situation can be yet more important.

However there have to be qualifications. The broad range of the climate only induces a broad pattern in the vegetation; the soils with their differing elements, the microclimatology, and the landscape all distract in their own fashion. Nevertheless, as well as a basic tree-line, which defines the traditional limit of trees, there is also a maximum tree-line, where no trees can survive however much the soil, landscape and microclimate might be in their favour because the overall climate is just too severe. Every species, except ubiquitous man, can only survive within a certain range of climate. Away from that range it will not thrive; the line will have been drawn.

Yet life of a sort lives almost everywhere. It survives the days, and it survives the years. It exists at the poles, in the tropics, in the temperate zones between them, and in the timelessness of the caves. These four situations are the subject of the next four sections of this book.

8 The Polar Year

A human story provides a suitable introduction to the polar world. Most humans do not live in such freezing circumstances, and find it hard to imagine extremely low temperatures. Peter Freuchen and Finn Salomonsen in their excellent book *The Arctic Year* tell a desperate tale. Some Eskimos had to make a journey in Baffin Land when no sledges were available. The journey could not be postponed, and on the day the sun returned to the sky the Eskimo party set off with makeshift sledges. These had caribou skins for the runners, soaked in water and then shaped by iceblocks, plus various crossbars. Wood of any kind was unavailable, and these cross members had been fashioned from slices of meat and from salmon; both had been frozen by the cold and both had been axed into the correct length and shape. Finally, the men used urine to help the skin runners travel more easily over the snow. As a model of improvisation, and of determination not to be outdone by the lack of proper facilities, the story thus far is hard to beat.

Unfortunately, on a subsequent night and when safely housed within a snow-block igloo, a Föhn wind arrived. Such a wind (thought to be named after Favonius, the Latin for west wind) is warm; not through any inherent warmth, but because there is such a pressure of cold wind descending from the mountains that the air below is heated by all this pressure. The effect of a Föhn wind below the mountains is a temporary warmth, a temporary raising of the temperature sometimes even to points above freezing. Such a momentary warmth happened to the meat-sledge party. They awoke to discover an unfamiliar climate, and looked at once to the sledges. Sure enough the dogs, always outside, always hungry and ready for an edible opportunity, had pounced upon the thawing bars and runners. The situation was therefore critical, even by Eskimo standards, and the party stayed where it was to wait for help. Even they considered further improvisation was impossible.

Soon they ate the dogs, they ate their clothes and, when help did come, only two women were left alive. A few gnawed bones were all that remained of the others. 'Have people been eating people?' asked the rescue team. 'We don't know,' said the women as they ate food, one cautiously and the other less so. It paid to be cautious, for the other woman died, and the cautious one lived to re-marry.

Eventually she produced four more children, and had thereby replaced her four children who had been with the party when the Föhn wind struck. Having restored the situation she said: 'Now I have a new husband and with him I have four new children, so I don't owe anything to anybody.'

Knowing, even in temperate climates, quite how unpleasant cold can be, it says something for human fortitude that the Arctic provides any kind of a home. Animals must also dislike cold. Perhaps this is anthropomorphism, but a human being requires slim intelligence and no percipience to know that the effect of a cold day upon his face, his ears, his toes and fingers is painful even in moderate doses. Severe cold is severely painful. Even so, neither man, nor other warm mammals, nor birds have totally avoided cold climates.

A mammal, according to a report by Laurence Irving in the *Scientific American*, must either generate ten times more heat in the Arctic than in the tropics, or clothe itself in ten times more effective insulation. Usually, of course, as with the Eskimos, the end result is a compromise; they both clothe themselves thickly and eat large quantities of fatty food, However the struggle for survival in the Arctic is so continuous, with the seals having to break open their freezing air-holes, and with the hungry bears patrolling the icy emptiness, that the more temperate latitudes might be thought to be more attractive, despite the numbers already living in them, despite the extra competition.

As an example of perversity there is Ross's gull which nests by the Kolima river (and elsewhere) and then goes *north* to Spitzbergen and the polar basin for the winter. The Kolima river, deep in Siberia at longitude 160° E, and within the Arctic circle, is still no picnic spot even in summer; but perhaps it seems so to a bird which then flies north and west to Spitzbergen as the days shorten to nothing at all, and as yet another winter settles over the land. The delights of the Kolima river in summer were well described by Douglas Botting (in *One Chilly Siberian Morning)* when he and John Bayliss arrived there in early June 1963.

The river then was still frozen, the ocean-going tankers caught by the ice in the previous September were still firmly locked, the people had neither buried anything nor anyone since the summer, and food stocks were low. In the past they would have been eating entrails, leather leggings and fishbones by this time. The dogs would have been eating each other. An eye-witness of those former days said: 'Everywhere one saw famine-swollen, livid faces, with fever-bright eyes from which despair looked out.' Winter, wrote Botting in more modern times, still passes through the area like a war; but all is changed when the river melts. It booms. It roars. The ice is ripped apart. 'She is moving, our little mother Kolima,' say the people, as the pancakes of ice first drift north to the sea. Simultaneously, or so it seemed to the two English visitors, 'the smell swept in like a plague. Normally, and despite the lack of sewerage, everything was kept as fresh as laundered linen by the cold; but in the thaw . . . !' Anyway, into three months the people compress a year's activity. They emerge, as it were, from a sleepless hibernation. They start to live again. Then, when everywhere else is thinking about autumn, winter descends once more. A fresh consignment of ships is locked up. The evenings draw in, rapidly. The nights lengthen and freeze. And the Ross's gull heads north for its own very particular brand of martyrdom.

Both the Kolima river and Baffin Land are within the Arctic circle. People do live there, somehow, but the situation within the Antarctic circle has always been different. No one, before the fairly recent exploratory and scientific invasion, lived in Antarctica, and even the current scientists hardly look upon the place as their home. Presumably the elementary reason for the earlier emptiness is that there is less land nearby to encourage people to go farther and farther south. To leap from Tierra del Fuego (which itself is an inhospitable area) over the sea to Graham Land and to Antarctica is asking a lot even of people who, farther north, make sledges out of frozen caribou meat. To go, or to be forced by rival communities, farther and farther north through the stepping-stone wilderness of northern Canada is more amenable than the big leap south from South America. The southernmost tips of Africa and Australasia are even farther north. (The most southerly point of New Zealand is near latitude 47° S, equivalent to Seattle and Lake Superior or St Nazaire and Dijon.) Cape Horn is only latitude 56° S, the same as Edinburgh or Copenhagen in the northern hemisphere. Admittedly the most northerly parts of the Arctic circle consist of no land whatsoever, but the Arctic circle embraces most of Greenland and large portions of the Asian and American land masses.

Consequently it is simple in the northern hemisphere for life to spread north overland until it can bear the conditions no longer, or to advance in summer and discreetly retreat when the savagery of winter is on its way. Man never even set foot on the Antarctic continent until 1895, when a Norwegian expedition landed at Cape Adare, and no one spent a winter there until a Belgian expedition did so three years later. If ubiquitous man took so long to get there it seems entirely reasonable that animal and plant life is thin on the ground. There is not even a single land-based mammal living on the highest, coldest and most barren continent of them all. Antarctica possesses 90% of the world's frozen water (most of the rest sits on Greenland), it adds up to a large proportion of the world's surface and yet there are so few inhabitants of any kind. The mammals to be seen on Antarctica are the seals, which are water-based, while the birds are predominantly penguins. They too, despite spending so much time on land, could be said to be water-based in that they feed there and swim there, but only two of the world's eighteen penguin species (the Adélie and the Emperor) actually breed on Antarctica.

The Arctic possesses men, polar bears, foxes, hares, caribou, wolves, musk-ox, lemmings, stoats (which are all land animals, even though the polar bear is a good amphibian), seals and walruses, and – notably in the summer – very many species of birds. Oddly there are no penguins, and even attempts to introduce them artificially have been unsuccessful. The penguin niche in the Arctic was most nearly filled by the great auk (which mankind battered into extinction in the early decades of the nineteenth century), and there are many other seabirds which lead comparable existences, but it is strange that there are no penguins, particularly when one species is found even on the equator in that archipelago of freakish zoology, the Galapagos Islands.

Seasonally both poles are identical in that the nights are equally long, the sun reappears – for each given latitude – at the same time of year (taking the six-month differential into account) and never climbs higher in the sky than 47° at the polar circles and 23½° at the poles themselves; but there are differences. Firstly, the frozen area of the Antarctic is some three times greater than that of the Arctic, thereby having a correspondingly larger influence upon climate as a whole. The Antarctic has the extra misfortune of being farther from the sun during its winter, whereas the Arctic has its winter somewhat mollified by being three million miles nearer.

Also Antarctica is higher, and therefore colder. On its coast the minimum temperatures in winter seldom fall below −70° F (−57° C), while the Russian Antarctic station Vostok, both inland and 13,000 ft up, has the distinction of being the world's coldest place. On 24 August 1960 its thermometer fell to −127° F (−88·3° C). The south pole itself, which Captain R. F. Scott had every reason to find disagreeable even in late summer, is 10,000 ft above sea-level. Its minimum temperature in winter varies between −79° F (−62° C) in the warm years to −101° F (−74° C) in the cold. Conversely, the north pole is at sea-level and warmer. It is better for both man and beast, whether in summer or winter, when they get there. 'Great God! This is an awful place,' said Scott at the south pole on 17 January 1912, and the Antarctic killed him some ten weeks later. The Arctic is relatively welcoming.

It is small comfort in either place that the regions are comparable in their precipitation to the hot deserts. Severely cold air is incapable of carrying much water vapour, and evaporation from the frozen and freezing seas is very slight. An average year in the Antarctic sees the equivalent of six to eight inches of water as precipitation, most of it as snow. In the Arctic the maximum is about twenty inches, but at Thule in northern Greenland it is about three inches, and at Ellesmere Island less than two inches. Many large deserts have far more.

Climatically the polar world can be defined as a region with perpetual frost, and with no significant permanent plant or animal life. The neighbouring region, slightly less bleak, is known as tundra, and is subjected to no month warmer than 50° F (10° C). The third cold region, less bleak even than the (probably) tree-less tundra, is customarily classified as the cold-temperate climate: it has at least six months with average temperatures below 43° F (6° C). As a general statement, albeit with exceptions (as with virtually every biological statement), the cold-temperate climate leads *only* to coniferous forest. Anywhere south of that region, and warmer, possesses – probably – deciduous trees as well.

Seasonally the equinoxes of the polar world are more crucial dates than the solstices. As with temperate regions the minimum temperatures are likely to be recorded after the winter solstice, but the polar temperatures wait until the equinox before they begin to rise significantly. At Vostok, that cold Russian base, the coldest days tend to occur a day or two after the sun's reappearance. This happens on 22 August, two months after the solstice, and the world's minimum temperature was recorded there two days later (in 1960). There is a similar time lag in the summer season when the relative warmth continues until a time near the autumn equinox. Of course, nowhere on Earth has its maximum heat on the midsummer day of the solstice, but the tropics come nearest to it (the rainy season permitting), the temperate zones less near, and the polar regions least near of all.

Nowhere within either polar circle is the seasonal range particularly extreme. This, so far as Antarctica is concerned, is because its summers are never very warm. The Arctic summers are also not warm and its winters are less cold than Antarctica; therefore its seasonal range is even more modest. The Arctic can never be really cold mainly because its sea-level ice-sheet sits on a (relatively) warm bath of sea-water at 29° F (1·6° C), and all this floating ice is never thick. It can be a few feet deep by May, but it will tend to melt during the summer. The parts that have neither drifted south nor melted before the subsequent winter will give a start to the next season's glaciation. Even so, the process is limited, and the maximum thickness of ice ever formed in the Arctic sea today is only some twelve to fifteen feet. Icebergs, which may be hundreds of feet thick, are of quite a different order: they can only come from glaciers, those frozen rivers which can either rush down-hill at 100 feet a day (Upernavik in summer) or scarcely move at all.

If an Eskimo kept a diary of one season, starting on 1 January, it would be begun in a time without sunshine. Nevertheless he could record seeing every midday the warm glow seen in the south, for nowhere does an Eskimo live too far north to miss this reminder of other people's daytime. In the so-called High Arctic, the northerly part, January has quite good weather, there is more light in the sky than in December while gales alternate with windless days. By 20 January, at Thule in northern Greenland, it is the time for seeing again the colour of your dogs. The seals, notably in Canada, can still be caught at their breathing holes, provided the catcher has sufficient clothing to withstand the vigilance necessary. (Corneas freezing over can be a hazard.) Female polar bears have mostly retired for a wintry sleep (different from hibernation) but the males may continue to hunt, mainly for seals. The females spend more time in their lairs, particularly those with young. Birds are rare at this time, most species having departed farther south. Some, such as the snowy owl, the ptarmigan, the raven and Hornemann's redpoll, are outstandingly hardy, but even they had moved south when winter approached.

February is cold, it is famous for its gales, there is usually more snow covering the ground than at any other time, and the unwelcome hot winds (which softened the improvisation mentioned at the start of this chapter) are an additional unpleasantness; but it is the time when the sun returns, and therefore much is forgiven. No longer causing just a diffused glow, the sun itself reappears to shine as a gently warming relief from the darkness of winter. The hot spells are unwelcome not just because they melt things, but because they are so rapidly followed by intense frosts. Clothes left outside the igloo soften in the warmth, become wet, freeze again into hopeless shapes, and are then impossible to put on. Rain may fall on the animals, as on the sheep of southern Greenland, then freeze into ice to fix the wretched animals where they were resting. Or, worst of all for the grazing caribou, fallen rain freezes on the grass to make it inaccessible. February is still very cold and fogs are common when the wind opens up patches of the relatively warm sea; animals are thin and hungry, and hunting them can be unrewarding, but the sun has returned as visible proof that the winter cannot go on for ever.

March sees the sun climbing higher and higher. People move more. The animals emerge. The young bears are still in their lairs, probably born in January. Animals are mostly thin, and hunters know that the toes of a thin bear conveniently point in, and of a fat bear point out. Thin meat does not taste good, but the fur is at its best. Many animals, including the bear and some seals, come in heat in March. No one from a temperate climate, feeling the cold of the polar March, and watching the sun climb so modestly every day, would call the occasion springtime. Nevertheless, March possesses all the virtues of spring–lengthening days, burgeoning offspring, warmer climates. The lemmings, for example, due to produce a litter a month throughout the summer, waste no time and often have their first brood in March. Carnivores are grateful for this prompt addition to the slender supplies of available food.

April is still unlike the temperate idea of spring, in that no leaf, no flower and no insect is yet to be seen in the northern Arctic; but the bird migration has started northward in optimistic anticipation. Most important of all, and often paying a penalty for their haste by arriving too early for food, the snow buntings have generally reached the most northerly lands of all by the end of April. Other birds, notably the great flocks of sea birds, move north with more caution, and are usually still south of their nesting sites during April. The fulmar is reckoned to be the first bird to reach its nesting areas in the northern Arctic, and waits until May before settling down at the particular cliff of its choice. In the more southerly Arctic the raven actually starts laying eggs in April. The sea eagle may also start in April in the south, but not until the end of May in the north. It is still cold, and a neglected nest will lead very quickly to a useless clutch of eggs.

April is a welcome month, when the people meet in great reunions. It was only in February that the sun reappeared, but by the end of April in the upper Arctic there is scarcely any night, let alone twenty-four hours of it. Snowblindness is an April danger, coming hard on the heels of the danger of seeing so little in the winter's dark. The caribou begin their migration, and leave their winter quarters. The ground squirrel, the only hibernator of the Arctic, wakes up. Most of the animals which did not mate in March do so this month. The bear cubs accompany their mothers. Bear skins are no longer much good. Harp seal pups lie about on the ice in the southern areas – they cannot swim for three weeks – and April is the best month for the seal hunters. It is still too cold to moult any fur. The first visit to the north pole, by Robert Peary in 1909, was in April.

May sees the end of winter. (It has been said there are only three seasons in the Arctic – winter, autumn and summer.) It is moving time for the Eskimos. Each day is now twenty-four hours of sunlight in the farthest north and almost so farther south; everywhere the sunshine is brilliant. Colours used to be invisible because of the darkness; now they are invisible because of the whiteness. Days can be above freezing, nights not so, but the daylight compensates for everything. Musk-ox calves are born in May and mammals are everywhere earlier in producing their young than the birds. The snow bunting, that pioneer among song birds, is still only building its nest in May, and only very rarely does it lay eggs before June.

May is a great month for bird song and the final rush of migration. Very few birds arrive at their nesting sites as late as mid-June. The snow is rapidly retreating in the south, although the dead hand of winter still sits on the north with frozen lakes and a frozen coastline. The impatience of migrants, who have such a short breeding season ahead of them, is entirely reasonable, but the reluctance of winter to leave their breeding areas causes a fearful toll. More birds die in May than at any other time. May is also the month of the great flowering of the phytoplankton which is nourished by the sunlight, and of the zooplankton which thrives on the phytoplankton. So the fish migrate to consume this food, and the seals migrate to consume the fish. This food chain snaps all its links together with amazing rapidity once May begins.

In May in the upper Arctic, plants on land are just about to flourish, with one or two exceptions. Insects are slightly earlier, appearing by mid-May. In the southern Arctic, May is full of flowers and insects. Animal traps are now useless because there is so much food about. The weather is warmer in May than in winter, but often less pleasant; the Arctic summer can mean long days of rain, unhappily compared with the colder but dry winter months. Elsewhere snowstorms are a frequent accompaniment of the May warmth. Moulting begins. No one from elsewhere would call the Arctic May a hot month, but the seals pant and flap their flippers to cool off and, should they fail to make a cooling plunge, may even die from heat-exhaustion.

June is the northern flower month. There is a suddenness about this flowering, from dormancy to full flowers, which is more like the desert's abrupt efflorescence than anything else in the temperate world. There can also be a tremendous outburst of insects. The mosquitoes of Canada are fearsome, and the butterflies of the south are more pleasantly unforgettable. Birds' eggs begin to hatch and most are hatched before the end of June, although the northernmost eiders are still only preparing their nests and laying. The Atlantic whales come north, the streams melt the ice clinging to fjords and inlets, the lakes melt, and each relaxation of the hostile environment enables more birds to fly in. Mammals multiply, birds invade in their tens of thousands, and the popular cliffs are alive in a way almost unthinkable a couple of months before. June is the time of the summer solstice; so the sun shines night and day and nature makes up for lost time. As the weather warms, the northern Eskimos move from ice igloos to tent-covered igloos, and by June they are all in tents.

July is proper summer. Small life abounds, because the algae abound, and everything that can do so takes advantage of the feast. The whales do well, the fish do well, and the birds live abundantly off the fish. There is still ice on the sea and in the fjords, and it is still possible to sledge on it, but the winds and currents move the floes this way and that. Consequently navigation can still be a hazard, and ships must beware being forced into situations surrounded by ice. The musk-ox leave their shaggy moulted hair all over the place. Mosquitoes and other insects are plentiful, although the number of different species in the Arctic is very low. Lower still are the reptiles with no representatives, and the amphibia with just a few near the Arctic circle border. In general the birds are busy feeding their young – August will be a frostier month. Some birds, notably the female phalaropes, leave their breeding places before the end of July, having spent a scant six weeks in the area, and the young come south with the males later on. July speeds up the glaciers, and is a good month for the formation of icebergs. The streams are all full, as if aware of the urgency and speed of the Arctic summer.

August is two-sided. Ships can travel farther north than in any other month, but the birds are almost all flying south. Exceptions are the fulmars; they have chosen their breeding spots in leisurely fashion, and with similar lack of haste they continue to cling to their choice. The young are on the cliff even in September, no longer fed by their parents but living temporarily on the fat they built up during the summer weeks beforehand. All birds end the summer with a lot of fat about them. August is still warm, it can rain a lot, but the nights are shrinking rapidly and everyone knows that winter is coming on fast. Flowers still flourish, but autumn is to be short, although never as abrupt as spring had been. Moulting, which has been going on irregularly throughout the summer, is being rapidly concluded. It can be a bad time for the geese, because moulting adults cannot fly and become easy prey. In the temperate regions to the south this is a hot month with all the gentle splendour of autumn to come. In the Arctic it is the beginning of the end, with a sense of hurry about it which is all-embracing; save, of course, for those leisurely fulmars.

September brings relief from the glare of summer. It is the month of the equinox, when both day and night are twelve hours long, and the sun rises less and less in the sky. The frosts kill the plant life, and winter has arrived before the end of the month. Lakes freeze over rapidly, and the sea less so. Snow reappears on the high places. The small streams are stopped in their tracks. The caribou now start to mate, whereas the seals and bears did so at the start of the summer. Soon, without birds, without insects, and with withered plants, the Arctic becomes again a dead land, all ready for winter. Some plants die colourfully, as in the autumn of the temperate world, but this polar death is far more abrupt. Animals that shroud themselves in white for the winter, such as foxes, stoats, lemmings and hares, do so in September. The northern Eskimos still live in tents, not yet in their igloos, although the southerners tend to live in tents nowadays all the year round. It is a time of making clothes for the winter. Animal pelts are good again. Strangely, according to Freuchen and Salomonsen, the departure of the sun and its nightly sinking below the horizon is greeted with enthusiasm by the very same people who applauded its reappearance only seven months before. An eternal blessing of the seasonal system is its constant change, and this is welcomed in the north even when the change is the sun's epilogue, the shift from brilliant sunlight to months of frozen darkness.

In northern Greenland the sun disappears for the winter in mid-October. Only a month beforehand day had equalled night. Therefore the reduction in daylength is extraordinarily fast, roughly half an hour every day. Temperate people have to tell each other that the days are drawing in; no one in the Arctic could fail to be totally aware of the sun's hasty retreat. The ice now makes it hard to hunt with kayaks but is insufficiently complete for smooth travel with sledges. Polar bears seek out suitable dens, and ground squirrels burrow holes before the frozen earth makes further excavation impossible. The squirrels then settle down for their hibernation.

Two months earlier food had been abundant. Suddenly all has changed: it is wondrous how birds like the gulls manage to find anything to keep alive on. The first igloos are made, and gleefully entered. They can be warmer than any tent. All the birds reach their winter quarters in October: the haste of travelling north, finding a site, laying the eggs, hatching them, rearing the young, and then retreating south before the onset of winter, is now over. Apart from an almost universal preference for more southerly latitudes in winter (Ross's gull being so exceptional), the birds are not uniform in the way they fly south, or in their ultimate destination. The snow buntings and wheatears spend the summer together on the west coast of Greenland, and then winter on separate continents; the snow buntings fly down past Newfoundland to the Great Lakes, while the wheatears launch off from Cape Farewell to the British Isles and the mainland of Europe. Greenland's east coast buntings go either to Europe or deep into Russia. Anyway, the land they have left (and the early-bird snow buntings are the last land-based birds to leave) is no longer the land they knew. The waterways close, any trapped ships stay trapped, the sunsets are superb (even if the sun is no longer seen to rise), and winter falls upon the world.

November is winter. The snow and ice have come, the sun has gone, and animals of any kind are rarely seen. Of course it is a critical time, with the lack of food being more disastrous than the lack of warmth. Practically every creature has its den, and certainly every creature finds it a lean time, save for those rare occasions when some huge carcass of a whale is thrown up on the shore. In place of the sun there are brilliant displays of the aurora borealis which may, at times, be even brighter and cast sharper shadows than the light of the moon. The Eskimos, not to be outdone by the enormous presence of the Arctic winter, like visiting in November: the fortitude of these people is almost unbelievable. A reason for visiting, apart from sociability, is that preserved food stores are abundant at this time, the snows are still gentle, and the real cold has not yet begun. Of course these terms are all relative – the food's abundance, the snow's gentleness and the cold's reality would not be appreciated by any temperate human in similar circumstances.

Even December, the darkest month of the year, lit by the moon and those flickering northern lights, is not as cold as the first months of the new year. It is a time of relative calm, of clear skies, of good travelling conditions, of visiting traps, even of bear hunting and harpooning seals at their blow holes. Nevertheless, despite the will of the Eskimos, despite the glowing affection of travellers for the land of the midnight sun, the Arctic winter is a period of emptiness. The darkness dominates, the cold is everywhere, the land is desolate, and even the inner sea is shrouded in its blankets of snow and ice. Nothing will really happen now until the sun comes above the horizon once again. The year ends in silence, and it is just as dark as when our hypothetical Eskimo started his diary of a chain of events he knows too well ever to record.

In a sense the polar year is an extreme extension of the conditions which exist in the temperate zones of Earth, being merely colder, darker and more forbidding in winter. In another sense it is a different world, for it closes down with a near finality whereas the temperate world only slows down. Like a holiday camp that is totally eclipsed by a temperate winter, the long polar night seems equally desolate, equally empty and barren. Then, like that holiday camp, and as soon as the days lengthen, the whole place opens up stridently, intensively, and makes up for lost time. Plants erupt, insects mature, birds fly in, and the frozen desert suddenly becomes a going concern. The seasonal change is crucial. It is not just a shift in emphasis, but a total dislocation of the previous state. It is light not darkness: crowds not emptiness; summer not winter. It is a different place.

9 The Temperate Year

The temperate region is oddly named. This word indicating moderation does not suggest the extreme fluctuations of temperature which are encountered within the region. The tropics are far more moderate, in that man is undoubtedly a tropical animal and the tropical fluctuations, such as they are, are less gruelling to his system. Nevertheless, the term does exist, applying to the two zones between the polar regions and the single tropical belt, and it will continue to be used. The intemperate temperate region is here to stay.

Perhaps it makes more sense to consider the word temperate short for temperature, because this factor is the key to the region. The three main climatic zones of the earth – the tropical, temperate and polar zones – have their three distinct types of season: the tropical season is the absence or presence of rain; the temperate season the rise and fall in temperature; and the polar season the presence or absence of light. These three variables are each primarily associated with only one of the three regions, but vegetation needs warmth and light and rain to thrive. Therefore each of these three requirements is most positively involved with one of the Earth's three climatic regions.

Admittedly rainfall varies with the seasons in both polar and temperate regions, and daylength varies in both tropical and temperate areas, and the heat and cold of the tropics and the poles also oscillate with the time of year; but these fluctuations are less vital. The difference between the dark polar world of winter and its bright summer months is paramount. The difference between a time of rain and the months without it is equally stark. And so too is the difference between cold months and warm months, however regular the annual rainfall might be, however adequate the sunlight even at midwinter. Temperature is the key to the temperate region.

Nevertheless there is more than one kind of temperature, and the temperate region is sub-divided. The warm-temperate climate is in two belts lying immediately to the north and south of the central tropical belt. These temperate subtropical belts are often hot, but their principal virtue for plant and animal life is that they are never cold. By definition, the average temperature of their most wintry month does not fall below 43° F (6° C). This is still 11° F (6° C) above freezing, and the term warm-temperate is therefore entirely justifiable. In the northern hemisphere this belt takes in the extreme southern states of the United States, all of Africa north of the Sahara, much of the northern shore of the Mediterranean, and Iran, Afghanistan, and southern China. In the southern hemisphere it includes much of Chile, of Argentina, the southern tip of South Africa and the southern half of Australia. Broadly it lies between latitudes 30° and 40°, whether north or south.

The cool-temperate regions lie poleward of the warm-temperate belts. They include the northern half of the United States, the southern and populated part of Canada, all of Britain, and most of Europe, southern Russia, northern China and Japan. In the southern hemisphere hardly anywhere falls into this category because there is hardly any land in the appropriate latitudes, save for the southern tip of South America and the southern part of New Zealand's South Island. It is this cool-temperate region which possesses the seasonal change that, to the self-centred conceit of cool-temperate dwellers, most typifies the four words – spring, summer, autumn and winter. The definition of the cool-temperate region is that it has at least one month but not more than five months with an average temperature below 43° F (6° C). The reason behind this apparently arbitrary choice of temperature is that at least one month occurs (but not more than five) which arrests plant growth. Hence the true winter and the true autumn leading up to it; hence the true spring, the return of activity, and the full summer months which follow.

Polewards again, and colder than the cool-temperate region, lies the truly cold climate, sometimes called the cold-temperate region. This area, already mentioned in the previous chapter, is warmer than the treeless zone of the tundra, but distinctly colder than the cool zone. It has at least six months that are colder, on average, than 43° F (6° C), and it has, in consequence, no deciduous trees, only the everlasting conifers. The great territorial mass of Siberia is predominantly cold-temperate; so too a large portion of Canada. Neither area has much population, and the coniferous forests lord it over the land. The areas may be hot in summer, as in eastern Siberia, but both the long spell of the period without growth and the weakness of the sunlight when it is at its strongest, prevent the areas from having either the potentiality of the warmer zones or the tremendous biological outburst of the polar summer. Sixty-two degrees, that latitude of minimum insolation where the sun is never high and where the days are never outstandingly long, runs through this cold-temperate region. Most of Scandinavia lies to the north of 62° N, but most Scandinavians live to the south of it. The northern part of Scandinavia is definitely cold-temperate, while the southern tips of Norway, Sweden and Finland lie within the cool-temperate zone.

It is amazing that any community exists within the cold areas of the Earth, and that anyone works during the heat of a tropical day; but it seems entirely understandable that most of the great civilizations of mankind have started within the temperate zone, and at the warm end of it. For human beings to thrive most beneficially life cannot be too difficult, as at the poles, or too enervating, as in the tropics. There must be some challenge, but not too much; the temperate seasonal oscillations create such a situation. W. G. Kendrew, the climatologist, concluded that 'seasonal change seems to be favourable both directly and indirectly to the highest human development'. Mesopotamia and the Mediterranean fit conveniently into both the warm-temperate zone and this argument, and they undoubtedly nurtured great civilizations. India, with all its civilizations, is tropical, but the rainy season is a most positive seasonal change. China is so large that it has many kinds of climate and change. 'At the equator,' added Kendrew,

> the monotony seems to undermine any civilizations that may be introduced. The most advanced civilizations have evolved in lands where the year goes round with a strong, but not excessive, seasonal rhythm punctuated by considerable weather changes.

Into this category fit the world's temperate places.

Firstly, is the warm-temperate zone, the most balmy of the three temperate areas. It is the warmest but its average annual temperature range can be quite considerable. Near the sea it is likely to be least; inland it is likely to be greatest. Strangely, the average daily fluctuation is sometimes greater than the average yearly fluctuation. This state of affairs is most likely to happen in a desert region, where the temperature is very hot at midday. In the middle of the night the clear sky, plus the dry and barren earth, jointly see to it that much of the heat is dissipated. Such conditions can produce a daily fluctuation from 60° F to 100° F (33–55° C), whereas the average yearly temperature range is probably less than 40° F (22° C) between summer and winter.

The dominant characteristic of this warm-temperate zone is that there is no winter pause in plant growth. On the other hand there may be a growth pause in the summer because rain in this zone is largely a winter event. There are, of course, all the exceptions. In South Africa, Capetown has 45% of its rain in June, July and August (i.e. winter), while Port Elizabeth, farther up the African coast, has its rain reasonably distributed throughout the year and Durban, still farther north, has only 9% of its annual rainfall in the three winter months. As a generality, the farther one travels towards the pole in the temperate zone the less positive is its dry season, whereas its cold winters become increasingly the limiting factor of growth.

Plant adaptation in the warm-temperate zone is more a matter of resisting drought than resisting cold. Consequently the short-rooted plants are less likely to flourish, for there is need of a long root to tap the deeper and moister regions of the soil in the dry period. In those places where grasses might have been expected there are shrubs and plants with fairly long roots. Frequently there is no more than bare earth between each plant, an unlikely situation in the cool-temperate world. These long-rooted plants are also resistant to water loss through transpiration, having thick, waxy and often hairy skins, while the leaves may be so rounded as to become spines. (Northern Europeans, accustomed to the gentle grass of summer, and to soft picnic spots in the meadows, are often taken aback by the unyielding harshness of the plant growth nearer the Mediterranean.) Also plants in the warm-temperate zones frequently hurry through their annual cycle and become dry seeds before the inevitable period of summer drought.

Most places in these warm-temperate climates do get a decent amount of rain in the year, say twenty inches or more, but the summer drought is shown up by the number of months whose rainfall does not add up to one inch. Gibraltar, for example, has three consecutive months without a total inch, Athens and Palermo have two, San Francisco has four, Jerusalem has six and Alexandria has eight. The yearly rainfalls of these six dry places are 35, 15, 25, 22, 24 and 8 inches respectively. These annual amounts form satisfactory rainfalls, save for the desert-like Alexandria, and all but Alexandria are greater than the average 20-inch rainfall of eastern England; but eastern England has rain all the year round, and slightly more in summer than in winter. The warm-temperate places have least when it is hot, and this difference is fundamental to the vegetation differences and therefore to the biology in general.

Obviously, however much climatologists may want to put them in, there are no firm lines to be drawn between different climates. Consequently, despite the firmness of the defined division between the tropics, and between the warm-temperate and cool-temperate zones, it comes as no surprise to those of us who did not divide the climatic world to discover that the warm-temperate has a bit of the tropics about it and a bit of the cool-temperate. Its rainfall is still a seasonal affair (like the tropics) but temperature is the dominant influence (like the rest of the temperate area).

However the cool-temperate climate could never be confused with the tropics. With that definition of from one to five months with an average temperature below 43° F (6° C), and always lying on the poleward side of latitude 40° in each hemisphere, and with a far wider range of temperature between summer and winter than any region nearer the equator, the cool-temperate zone is quite a different world. Its really cold season, its proper winter, and its positive cycle of agriculture are all most pronounced, and they become more so from west to east, from the sea to the interior, from Britain to Siberia, from Vancouver to Winnipeg, from Seattle to North Dakota. Consequently winter is treated more seriously the farther one advances into each continent. Within the British Isles, situated on the extreme westerly seaboard, there is always far less preparation for winter, and a modest snowfall is likely to disrupt many services. This is less true for the European continent or in much of the United States where winter is expected to happen once a year.

The cool-temperate zone is an area of westerly winds, which blow at all seasons but which are confused by the enormous numbers of depressions, the low-pressure areas. The westerly winds see to it that the westernmost portions of each land mass receive most wind and rain, but the depressions – more important in winter than summer – obfuscate the situation by causing the weather to come from all points of the compass. Warm and cold fronts constantly arise, as warm and cold air masses meet, to cause rapid changes in temperature. Weather prediction then goes awry. In warmer latitudes a shift in wind direction will still mean a flow of warm air because everywhere both to the north and south is warm. In the cool latitudes a shift will mean either cold air from the north or warm air from the south. Hence, the rapid changes of climate.

Deep within the American and Eurasian land masses the hottest and coldest times of the year occur usually soon after the two solstices of June and December. (The time lag is less than near the poles.) On the coastal areas, due to the reluctance of the sea first to warm up and then to cool down, the peak temperatures may be considerably retarded, say to August and February. In these places the hottest and coldest times may therefore be nearer the equinoxes than the solstices of midsummer and midwinter, and therefore similar in this respect to the Arctic and Antarctic.

Now to the cool-temperate year, and to the cycle of events as they impinge upon the climatically obsessed residents of the cool-temperate zone. Spring is a good time to start, not just because it is the restart of things but because spring is peculiar to the temperate world, and most of all to its cool-temperate zone. The tropical world certainly has a resurgence of life when the rain returns but it is not a spring, and the polar world has no time for such a leisurely reawakening. The three months of the temperate spring are March, April and May but snowdrop flowers and even a venturesome crocus can appear in January. The snowdrops can even be over in certain places by the end of February, the crocuses can be (and often are) in full bloom, many birds are cavorting before each other, while all bulbs have pushed shoots through the earth, and everything – in a warm year – seems to be on the go even before the three months of spring have begun. Then, as tree after tree begins to cover its bare limbs, first with small protuberances, then with an effusion of leaves, spring goes on and on. Who can say it is over until all the trees have their summer covering?

The reason for this lengthy prologue to the full bloom of summer is that the climate is dickering on the very brink of the growth temperature. Were winter to continue until, quite suddenly, the temperature of 43° F (6° C) was reached, and thereafter surpassed, spring would indeed be an abrupt affair. Plants would either not be growing or, suddenly, they would be active. The situation would be as in the polar regions where spring is non-existent and summer happens almost at once. The temperate spring has frosts, which put everything back, and warm days, which bring everything forward, and days in between when the temperature hovers unhelpfully between growth and non-growth. Here and there a sheltered spot may just tip the scales in favour of growth, while a few yards away a stream of cold air may forbid any such inclinations. One plant may find the existing conditions acceptable for growth; another may find them still unendurable. Obviously snowdrops, which are early, and bluebells, which are late, react differently to the contrary climate of the year's first months. So spring goes on and on, and from the first push of the first snowdrop to the last appearance of the last leaf may be nearer six months than three.

While the Arctic is still gripped most firmly by winter any wind coming from there during March or April, even if it curls round to approach the cool-temperate zone from the south-west, is still perishingly cold. On the other hand if the wind stops, and if the sun shines brightly through the clear sky, the spring days can suddenly become almost hot. April makes the thermometers in Britain rise to 70° F (21° C) roughly once in every two years. Winds coming from the tropics are cooler than in summer because the tropics are cool at this time. Consequently they carry less water vapour than normal, and therefore the spring months can be very dry. April showers may be irritating, but both March and April are generally drier months than January and February. An unfortunate and crucial factor of the various temperate places is the custom of at least one or two frosty nights in May; the beneficial effects of the March and April warmth can be snuffed out in a single night, and fruit growers try every means to prevent these annual reminders that their plantations are not a few degrees of latitude farther south.

Everything in the cool-temperate animal world happens earlier than within the Arctic circle. The winter migrants are still around at the start of February, but moving more and more by the end of the month. Courtship of many bird species begins even in the winter, birdsong increases during February, and some birds, such as the raven and heron, look to their nests. Hibernating animals are still asleep in January and February, but wake up with an unevenness much like that of the sprouting of bulbs. Generalizations about springtime events are peculiarly difficult but it is safe to say that animals are being as leisurely and uncertain about the onset of summer as are the plants.

March is really the time of the great awakening. The hibernators warm up, the mad hares leap about in masculine rivalry, and the migrants start arriving at the very end of the month. When they reach their destination they often find the local residents have stolen a march on them, for most resident birds have started nesting by the time the first migrants arrive. The amphibians lay their eggs in March and April, the reptiles come out to bask in any available sunshine, the fish start to spawn (although trout can be far earlier), and the insects start to flourish in their myriad fashion. Even so, despite all this activity, there is no sudden emergence comparable with the polar world. There is more time in the temperate zone, and life takes more time in consequence.

By May all the bats are out, their young are born either then or in June, and so are the young of many other mammals. May is also the month for ornithologists, as every bird is with young, or with eggs, or with a nest. Unlike the Arctic these clutches may only be the first of two or three, all of which can be reared in the longer summer. (In the Arctic there is no occasion for such repetitive dalliance.) May is also croaking month for the amphibians, but their egg-laying, as with the birds, can be prolonged even into August. Fish spawning gathers pace, and nearly all the river fish are laying their eggs in May or June. April has often been disconcerting, but by May the insects find the weather warm enough to multiply rapidly as further harbingers of the protracted spring. Their eggs hatch, their nymphs shed skins, their larvae become adult and the last of the adults emerge from their winter hiding.

The apparently insect-less world of winter then becomes the visibly insect-full world of summer as tens of thousands of insect species become active. And, of course, the botanical world comes into its own to take advantage of the summer sun. Some, such as the bulk of the woodland flowers, have to take premature advantage of it, before the leaf canopy spreads above them to make the woodland floor either as dark or darker than in winter. Consequently, in the fullness of summer there is about as little or as much plant growth beneath a beech tree canopy as there is on some windswept section of the northern tundra.

June is emphatically the start of a new season. Daytime air temperature is probably above 60° F (16° C). There is a feeling of warmth even on a cloudy day. The surface of the sea has lost much of its wintry chill, and the days are as long as they will ever be. High-pressure regions tend to settle over the Azores, and western Europe is likely to come within their good weather influence, making June the sunniest month of the year. The depressions still exist, they still move from west to east, but they are less deep and so the winds around them are less strong. However the temperate zone would still get its winds either from the tropics or the poles, fronts between two air masses are still the rule, and temperature changes are always to be expected. Summer rain is heavy when it comes, making the season one of greater rainfall but of fewer wet days. Thunderstorms are also a summer feature, often following a spell of fair weather. (That any indulgence leads to inevitable retribution is a belief firmly supported by the British climate.)

The heat of high summer, of July and August, itself adds to this retribution. The sea is warmer, and this leads both to wetter days and cloudier skies. Depressions in August, should they occur, can bring floods of rain and have a disastrous effect. Even without them the British climate in late summer is less dependable. Before grain drying there was always far more doubt about the harvest, and would the fine days stay fine? Nowadays it is not so much a matter of sprouting corn, but of a crop so flattened by wind and rain that the combine harvester cannot consume it. What with fruit-buds being lethally nipped by late frosts, with corn left damply in the fields, with potatoes ruined by cold weather, and with vegetables soggy with slugs and dampness, one wonders what on earth much of the cool-temperate zone is fit for – save for grass. Nowhere else can grass grow like it grows in the British Isles.

By September, a drier month than August, the migrant birds have probably left. It often seems in any one spot as if they have not yet departed because the migrants returning from farther north misleadingly suggest that the local birds have not yet flown. Also the actual departure of each species can be blurred by the return to Britain of other species which have spent the summer elsewhere. There is no total holiday-camp departure as in the Arctic. Instead some come and some go throughout the year: the cuckoo leaves and large numbers of ducks fly in.

September is a good month for caterpillars, for wasps, for flies, for spiders and for mites. The very abundance of countless species seems to suggest an eternal summer. Unlike the migrant birds, which call it a day and fly home, the residents continue to take advantage of the slow retreat of summer. The weeds are particularly adept at making use of autumn, and they can dominate this time. Weeds are not just plants which are growing where they should not, but plants which are quick to take advantage of bare ground, of the unnatural business of agriculture. They are pre-eminent in the late summer; and so too are mushrooms and most fungi.

With any luck the month of October (when Eskimos are having their last sight of the sun) is as nice or nicer than September in the temperate world. Indian summers, those most unhurried autumns of them all, can keep winter at bay apparently for ever while the leaves slowly brown, and the warm air blows lazily over the ground. On these occasions it is a quiet anticyclonic system doing all the good work. More frequently October is the month when there are sharp climatic clashes

The fungus Lactarius Torminosus is an example of vegetation that demands constant damp conditions. There are over fifty thousand types of fungus known, including mushrooms and penicillin.

The temperate zones The cool middle belt of the temperate zones (including most of Europe, North America and New Zealand) provides the best example of the seasonal cycle conforming to the words spring, summer, autumn, winter.
LEFT: England now that April's there. The English spring is gradual and prolonged, supposedly lasting from March to May, but in fact stretching from the first snowdrop in January to high summer when the last leaves on the trees come out.
ABOVE: Autumn in Vermont. Indian summers can last right through October, causing a lengthy period of colour changes in the leaves.

Summer and winter in the temperate zones show far greater temperature differences than in regions nearer the tropics, although both seasons are subject to variable weather.

LEFT: Summer in Sussex: a field of ripe corn waiting for the harvest. Although the British summer rarely turns out as badly as Charles II's 'three fine days and a thunderstorm', late August is unreliable, justifying the traditional uncertainty over the weather at harvest time.

ABOVE: Winter in the Westerwald, Germany. Winter increases in severity according to distance from the sea as well as from the equator, and the temperate zone can include extreme conditions.

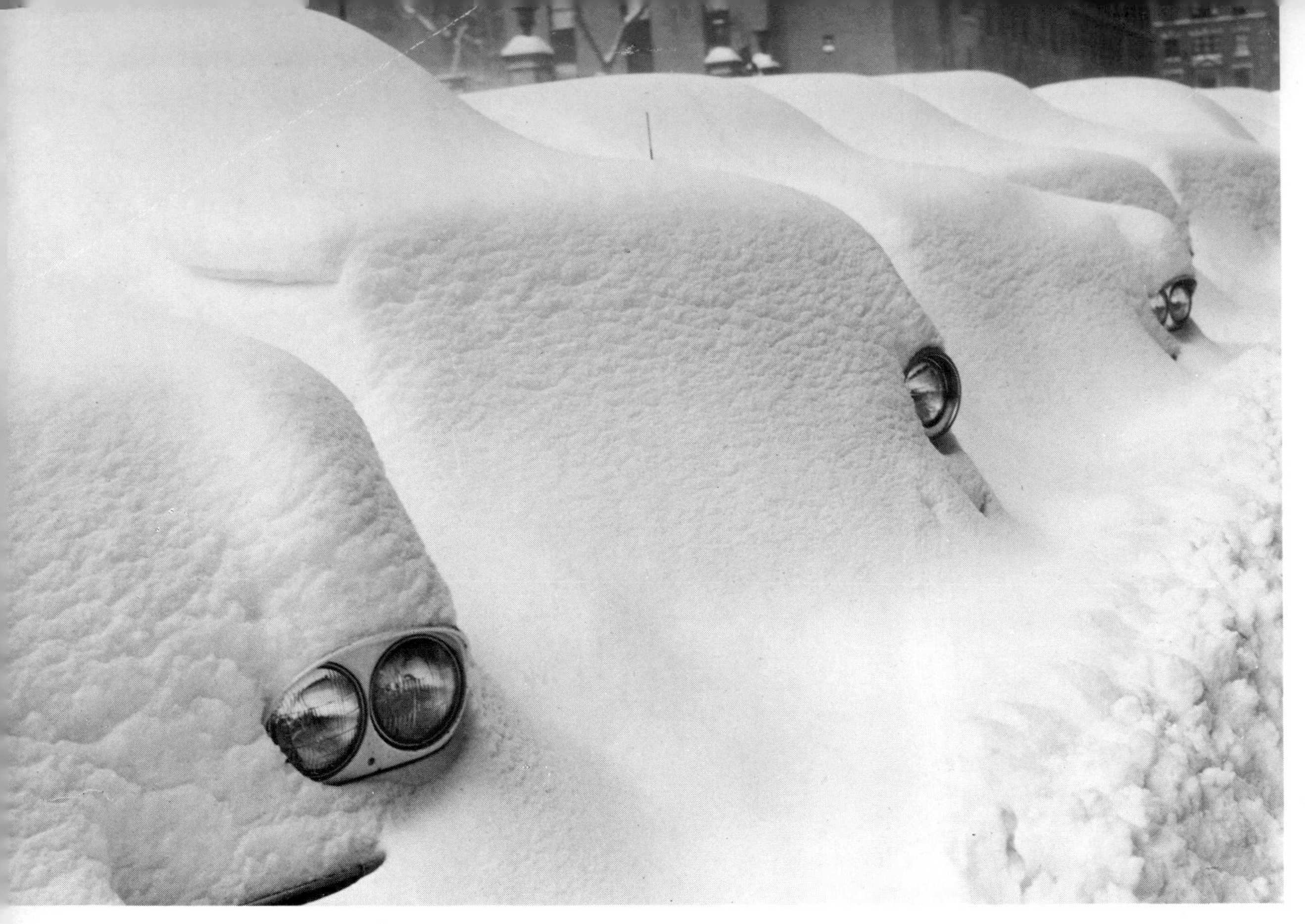

ABOVE: Winter scene in New York City. Although snow causes chaos to traffic, it performs the valuable function of insulating plant life from further heat loss.
BELOW: Ice on lakes and rivers similarly insulates the fish underneath against the cold. Freak conditions also have their compensations for humans: ice hockey on the Thames at Richmond in 1855.

Hail can occur anywhere, but the most damaging storms occur in warm temperate conditions.
ABOVE: Light aircraft after a hailstorm.

BELOW: A field of corn battered by hailstones in Indiana, U.S.A. Hailstones the size of hens' eggs have been reported.

Floods may cause considerable damage and loss of life, but they also deposit valuable minerals on the land.
ABOVE: Flooded countryside near Exeter.

BELOW: The Devon village of Lynmouth after the River Lyn had burst its banks in August 1952: thirty-one people died and nine hundred were left homeless.

The tropics The seasonal rhythm dominating the tropical year is the arrival and departure of the rains. The monsoon condition is most marked in India: during June the south-east winds shift in direction and come in from the Indian Ocean, putting a sudden and violent end to the drought—in Bombay the monsoon season from June to September produces seventy inches of rain compared with seventy millimetres for the rest of the year. Rice is a crop completely dependent on the monsoon.

BELOW: Before the rains.

RIGHT: After the rains: a paddy field.

If the monsoon rains come too late or too soon or stop prematurely, it may mean disaster for the rice crop and for the millions who rely on traditional timing and guesswork.
ABOVE: A village in India parched by the sun.
LEFT: Indians carrying their belongings from their flooded homes.

Not all tropical areas are subject to monsoon conditions: the majority of the world's deserts—where the problem is not rain but the lack of it—lie within 15° north and 40° south of the equator. Deserts are hostile to plant life, but surprisingly support roughly the same number of birds, mammals and reptiles as more humid areas. Most desert animals are sand-coloured, nocturnal, live in holes and have large feet.

ABOVE: A desert rodent, the jerbil.

RIGHT: The Australian stumpy-tailed lizard has adapted itself to drought and starvation conditions by storing fat in its tail.

BELOW: A freak fall of rain can turn a desert into a mass of bloom. Some seeds can survive the drought and flower annually or ephemerally depending on chance rainfall.

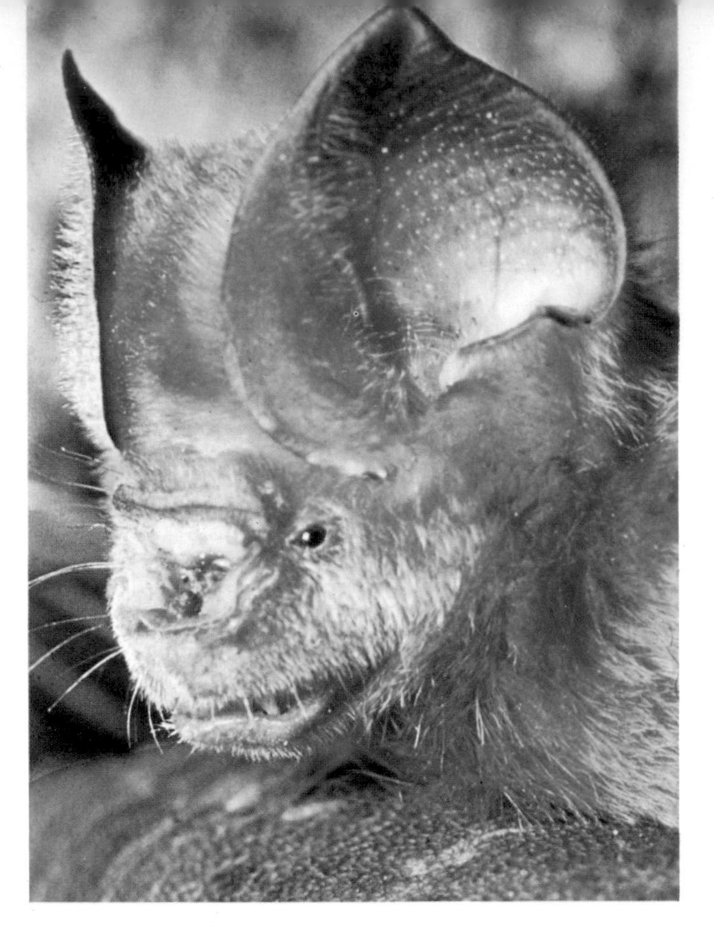

Plant and animal life thrives in the hot wet tropics. Many animal species associated with the temperate zones have a tropical counterpart.
LEFT: The long-eared bat, a tropical member of the bat family.

BELOW: The Greater Kudu, one of the largest of the antelope family, found around the Zambesi. The male keeps away from the female (shown here) except during the breeding season.

beween the Arctic (by now getting very cold) and the Atlantic (still warm). This conflict is always the cause of depressions, of whole sequences of low pressure areas, and of innumerable fronts – in short, of the familiar wind, rain and foul weather. When clear days come again they are probably anticyclonic, and partnered by the first frosts. The low temperature hastens the death of leaves, and their colouring can be superb. The colour change in western Europe is more protracted than farther north or in, say, Canada and New England where the arrival of cold is extremely savage, causing a quicker, firmer and more glorious response.

November and December are wintertime, although really cold weather is unlikely in November. The days are short, being shortest in December, and November is a reliable month in that neither hot nor very cold days are likely. December is more contrary; both warm and extremely cold days are possible. For animals the harsh time has come. Hibernators hibernate. Many insects exist either in a dormant state, or solely as females, or as eggs. The fish swim deeper, while the resident birds continue with their old way of life, and suffer accordingly. Large mammals do not store food, save for the fat on their own bodies. Small mammals, such as the mole and squirrel, frequently do so. Water-living animals have the great benefit of water's properties. It provides a more equable climate at all times, and in winter the surface ice insulates the water below it from becoming colder.

The odd fact about water is that as it cools it contracts in volume, until it reaches 39° F (4° C), and thereafter, unlike all those substances which continue to shrink with decreasing temperature, it expands. But ice, which finally results, weighs only 57·5 lbs a cubic foot (as against 62·3 lbs a cubic foot for fresh water at 70° F (21° C)). Consequently the warmer water is heavier, and sinks to the bottom of the pond. The ice above protects it, and any fish, amphibian or invertebrate can seek out the temperature layer most to its liking. When the thaw comes, and the ice melts, this oddity of water is again an asset. The surface water, warmed by the sun, grows heavier and sinks to the bottom mixing with the water which has lain down there since the freeze. It is hard to imagine quite how different the water world would be without this remarkable characteristic.

Another property of water is its ability to form snow, and then to lie around as snow. It thus covers the earth and the food supply, but it also insulates the earth from further heat loss. It is possible for snowflakes to reach the ground even when the thermometer is as high as 43° F (6° C) if conditions are right, i.e. if the air is not saturated, and if it is windy – as sometimes happens in April. Generally, a temperature of 38° F (3° C) on a day of saturated air, provided precipitation is occurring, will yield the occasional snowflake. At 36° F (2° C) there will be quite a few, and at 34° F (1° C) there will be a lot of large melting flakes. As a rule the lower the temperature the finer the snow.

Britain is a small country but quite large enough to demonstrate increase of snow with latitude. In the south, Cornwall has the lowest average snowfall with less than five days a year on which snow is seen to fall, the Midlands have about fifteen days, Edinburgh has twenty days, the Scottish Highlands have from twenty-five to thirty days, and the extreme north-easterly tip of the British Isles has about thirty-five days. On the other hand there are far fewer days when snow actually settles. Arbitrarily the climatologists have decided that if, within a certain radius, half the country surrounding a particular spot is covered with snow at 9 a.m. that particular day can be classified as snow-covered. Whether snow settles depends of course on temperature, and there is a correlation between the average temperature of the month and the number of snow-covered mornings. If the mean temperature is 42° F (5° C) there will be (perhaps) one day, if 40° F (4° C) then three days, if 38° F (3° C) five days, if 36° F (2° C) eight days, if 34° F (1° C) fifteen days, if 32° F (0° C) twenty-four days. In the Scottish Highlands there are substantial areas with snow lying for over a hundred days.

It is just possible in Britain foı the occasional drift of snow to survive the summer and still to exist when the first fresh snow falls to cover it, but this can only happen in the hills. Temperature decreases with altitude, and there is almost a linear relationship in Britain between snow-covered days and height above sea-level. In Scotland there is an extra day of snow cover per year for every twenty-four feet rise in altitude. Upper Scotland (above 2,000 ft) has the ground covered by snow for as long as the low Alpine resorts ten degrees of latitude farther south. Britain would always have snow on its mountain peaks, according to Professor Gordon Manley, if the highest mountain, Ben Nevis, exceeded 5,300 ft, if the Lake District topped 5,900 ft and if the mountains of North Wales were above 6,300 ft. Therefore these three areas are roughly 900 ft, 2,700 ft, and 2,800 ft too low to have permanent snow. The mountains naturally have to be taller in the more southerly latitudes, and for this reason equatorial Mt Kilimanjaro only just has its permanent snow cap, despite its peak being an impressive 19,000 ft above sea-level.

By mid-January, in the traditional depths of winter, depressions cavort over the British Isles in a steady fashion. They cause a lot of cloud, and this means an average daytime warmth of 46° F (8° C), and a modest night-time fall to 38° F (3° C). Most of January is above freezing in England, even though it is the coldest month. January 1940 was a cold year, well remembered for skating and freezing cold for days on end, but its mean temperature – at Cambridge – was only 29·3° F (−1·5° C). January 1916 was exceptional in the other direction because the thermometer – again at Cambridge – only dropped below freezing on four nights in the month; its mean temperature was 44·5° F (7° C), or warm enough to start plants growing.

January, like practically every climatic event, is not always true to form. Drizzle is frequent, snow is always likely, but frozen snow on the ground in most parts is improbable. By February the sea is colder and polar winds can be exceptionally severe, but the days are lengthening and the month is warmer than January. Human beings may not notice the fact, feeling this month to be just as unpleasant as its predecessor, but the plants notice the increasing daylength, and spring slowly emerges in its own gentle fashion. Wintertime, possibly severe but probably not, slowly yields its hold upon the country and everything, once more, is reborn. The perennial cycle starts again.

With snow in April, frosts in May, bitingly cold nights in midsummer, and balmy days in December and January, it might seem as if the seasons were some kind of myth, with an archaic response to them maintained stubbornly by the trees as they shed their leaves, and by other organisms as they undergo their age-old processes. Given half a chance, and a good year, winter scarcely exists, save as the occasional cold snap. Given half a chance, and a bad year, summer hardly exists either, save for King Charles II's three fine days followed by a thunderstorm – his description of a British summer. Every month does indeed have days which could well have been taken from any other month in the year, and the reason is the location of this temperate zone, between the devil and the deep blue sea of the tropics and the poles. The Earth's rotation sets up a basic pattern of prevailing westerly weather but the hot and cold spots to the north and south cause high and low pressures which are further confused by the highs and lows caused by the continents.

So the temperate picture is of a seasonal variation, modified by great draughts of cold air, sometimes coming over a cold sea, sometimes over a warm sea, and then by large douches of warm air, sometimes very wet, sometimes less so. There is rain all the year round (except in the continental centres) and near the sea there is likely to be coldness and warmth all the year round. There is still an average temperature differential between summer and winter – 11° F (6° C) at San Francisco, 20° F (11° C) in Britain, and over 100° F (55° C) in central Siberia. There is always this cycle, whatever its amplitude, and temperate life responds to it. The only zone composed strictly of four seasons does not have the stop-go pattern of the poles and, in consequence, its effects are more varied, possibly of greater significance in the history of evolution. By being so varied the temperate world encourages considerable variation in response to it. Natural selection then encourages or discourages each particular variant and evolution takes place. Probably, therefore, the temperate world has been of greater biological significance than the polar world, but probably the tropical world eclipses both of them in its importance.

10 The Tropical Year

The tropical world is the hot world, where the days are never short nor the nights ever long, where the daily sun shoots high in the sky and then drops abruptly down to Earth again. It has seasons, neither as the Eskimo would understand them (no black winter, no bright summer), nor as the temperate farmer understands them (no cold period calling a halt to growth), but there is, almost always, a definite climatic rhythm to the year. Even the word *monsoon* means season. Daylength does change, from thirteen and a half hours to ten and a half hours within the 47 latitude degrees of the tropics, and heat changes as the overhead sun oscillates over the line of the equator. But it is the arrival and departure of the wet season that is the major factor, and this is what the Indian is referring to when he uses the word *monsoon*.

Hot climates have been defined by two main classifications. One says they have average annual temperatures of at least 70° F (21° C); the other says the average temperature of the coldest month must be 64° F (18° C) (and therefore the hot climates are bounded in each hemisphere by this isotherm). There are cold times in the hot regions, notably in the desert areas at night, and frost can occur, but the daytime heat compensates for these drops and average temperatures are well above that all-important point in the scale, the 43° F (6° C) responsible for growth.

Essentially the hot climates are within an area slightly greater than the actual tropics. There is a simplicity about them. They are hottest because the sun is always high in the sky, and its radiation passes through the least distance of dissipating atmosphere. They are also farthest from the freezing poles, and are buffeted by the temperate zones against any inroads of chilling arctic air.

The high pressure region of the horse latitudes, roughly between 30° and 40°, forms the main barrier. (The name allegedly arose because it was the custom to throw horses overboard should a ship be becalmed for a long period. The horses would be drinking too much of the water for too long.) It is this region which leads both to the westerlies and to the trade winds; the westerlies are on the poleward side of the horse latitudes, while the trade winds are on the equator side. The horse latitudes themselves, which are spewing out these winds, are relatively calm and doldrums are common there. With the high-pressure horse latitudes of both the northern and the southern hemisphere dispatching the famous north-east and south-east trade winds towards the equator there is a further doldrum zone where these two wind forces meet to cancel each other out. This happens either at the equator itself or slightly to one side of it according to the season. The constancy and reliability of the trade winds, or trades, gives them their name, reflecting their suitability for all the trades making such regular use of them.

There are various possible sub-divisions of the 60 or so degrees of latitude which include the hot regions. The central and equatorial division is the doldrum area of the equator. This receives some rain all the year round. Then there is the so-called tropical division, itself divided into continental (with rain only in the hot season), and marine (with rain all the year round). Finally there is the desert division, also split into continental and marine; both have next to no rain, but the latter is colder. In all three the yearly movement of the sun is pre-eminent, both in its heat and in its manufacture of the rains, but nowhere is there such consistency of climate as in the equatorial band.

Ocean Island, one degree south of the equator in the Pacific, suffers an average temperature of 81° F (27° C) each and every month of every year. There is an annual rainfall of eighty-four inches with no month having more than eleven and a half inches (January) and no month less then five inches (June). Such monotony must be impressive. So too at Quito, right on the equator in Ecuador. It is high up (9,350 ft) and cool, but every month has an average of 55° F (13° C) save for November which has 54° F (12° C). Rainfall is forty-two inches in the year, with every month having at least one and no more than seven inches. Nairobi, 5,495 ft up in East Africa, and only one degree from the equator, seems almost turbulent by comparison. Its peak month is 66° F (19° C), its lowest is 58° F (15° C), and its rainfall varies from less than an inch per month up to eight inches. Some might consider its climate tedious and unvarying, but certainly no visitor would do so who came either from Ocean Island or Quito, Ecuador. Within the equatorial zone as a whole, and discounting the effects of altitude, the total average monthly temperature around the entire circumference of the globe throughout the year seldom tops 100° F (38° C) or falls below 60° F (15° C). It is a most uniform belt.

Such rises in the rainfall as do occur in the equatorial zone usually do so soon after the period when the sun has been overhead. For the equator the peak times are April and November but, although both periods of rain have similar causes, one of the two peaks is frequently greater than the other, leading to the big rains and the small rains. Within the tropics as a whole there are variations to this theme; the small rains may be totally suppressed or, farthest from the equator, both sets of rains may coalesce to form one rainy period. In general dry seasons follow the solstices. Unlike the temperate rainfall, which can come at any hour and for any one of a host of reasons, equatorial rain can be pleasantly predictable. Bright sky often starts the day, but clouds build up until, with or without thunder, the rain comes; thereafter the evening is quiet and reliably dry. People do not set their watches by the phenomenon but, in many places, they do feel they can count on it to perform correctly.

The Congo river is affected most conveniently by the peak rains; its branches lie to the north and south of the equator and it gains from the different rainy seasons of each hemisphere. Its northern tributaries receive heavy rain in May, and the southern ones in December. Both surges of water see to it that the main Congo river is well supplied, and navigable, all the year round. Normally the presence of a dry season has a profound effect upon river flow, and thus upon the transport, commerce and irrigation dependent upon it. The world's largest river system, the Amazon basin, is said to contain one fifth of the world's fresh water, and is not unduly subject to the seasons partly because it too experiences rainy periods on both sides of the meteorological equator and because its vast catchment area receives 3,000 cubic miles of water a year. This ensures a large supply of water to the main river, quite apart from the reliable trade winds always bringing moist air from the east. (Assuming England and Wales or the state of Pennsylvania to have 50,000 square miles, and assuming the average rainfall to be forty inches a year, the total rainfall in each place is therefore thirty-six cubic miles. The Amazon basin, is, as John Gunther wrote, the home of many of the world's superlatives.)

Because an equatorial climate has such an adequate supply of warmth, light and water all the year round, its vegetation responds to the monotony of having little overall pattern in tune with the year. Look at a rain forest with leaf shedding, fruiting, flowering and growing going on all about you. Look for the rings of a tree, and you will find them hard – if not impossible – to distinguish, being quite unlike the permanent reminders of winter in temperate trees.

Botanically (and zoologically) there are many more species than in the temperate region. It is possible to find as many plant species within two tropical square miles as exist throughout the length and breadth of the British Isles. (Prof. E. J. H. Corner of Cambridge University once proved this point by collecting well over a thousand plant types within a mile or two of an equatorial camp.) Tropical forests are not dominated by single species, as is the pattern in Europe or parts of the United States with beech or oak or birch often being the most numerous. It is hard finding two specimens of a single tropical species standing next to each other. Quite why there should be all this diversity no one knows.

Agriculturally the dry season tends to be the time of harvest, but there is less regimentation about this than in the temperate world. The very profusion of the tropical growth, and the speed of this growth, has led to a much more undisciplined agriculture than in the harsher temperate world. A West African who burns down a bit of forest, plants a few yams between the fallen trunks, and then moves on when the earth is exhausted, is being appallingly casual in the view of, say, the Dutch farmer, who extracts every ounce from the neat fields around him and who does not understand the tropical deficiencies. It is the burgeoning tropics which have encouraged, and which permit, such practice. And the seasonal year, bringing reliable warmth and rain, adds its blessing. (On the other hand the Dutch farmer has less to fear if he forgets a field for a while. It does not have the urgent tropical ability to become scrubland the moment his back is turned.)

Everything said about the tropical climate, with its equable conditions, becomes less and less true farther from the equator. Temperature is higher and there is greater daily and annual variation. The rains are more erratic. Droughts and frosts are possible. Consequently, harvest times become much more definite. Incidentally, European invaders of tropical America were confused by the seasons they encountered. Mediterranean winters are wetter than the summers. Therefore the Europeans persisted in calling the wet season *inverno* (or winter) and the dry season *verano* (or summer). In fact the tropical wet season is summertime, in that the sun is then higher in the sky, and the Spaniards and Portuguese should have realized they had arrived not only in a new world but in a new climatic zone.

The increase in unreliability farther from the equator can have disastrous effects. Not only can the rainfall vary but so can the manner of its arrival. The total variation is bad enough, but too much rain in too short a time can lead to extreme run-off when the water pours over the land and then goes straight into the rivers. Also, with a rainy season coming in summer, evaporation can be outstandingly high. A wet day in England can bring one-tenth of an inch of rain (this is the average figure for days on which rain is recorded) and the lack of evaporation can leave the ground extremely moist from this and other equally modest rainfalls. One-tenth of an inch in the tropics may be barely detectable an hour later. Thirty inches a year, an average figure for London or most of the area around the Great Lakes of mid-west America, is more than adequate for agriculture in a temperate world. It is marginal in the tropics.

Disaster comes when a marginal situation is relied upon by millions of people. Calcutta has a rainfall which varies annually by 16%, but Lahore's smaller rainfall varies by 38%. Consequently, as crops are chosen and planted according to the presumed rainfall of an average year, it is extremely likely that the crops will fail. A wise man would suggest greater caution or more reliance upon irrigation, but a country like India has traditionally hoped only for the best. Every so often it receives the worst. The 1967 lack of rains in Bihar, which caused such famine, was entirely normal in the sense that Bihar's rainfall is normally unreliable and years with 1967's rainfall had been amply recorded.

Deserts and monsoon zones are the extreme ends of the tropical situation. One has a dominant dry season, only occasionally interrupted by rain. The other has a rainy period, which either can be or can fail to be bountifully lavish. The deserts have few people; the monsoon areas have many. In one sense life is less hazardous in the deserts because the rain there is never trusted. The monsoons are trusted, and millions have died by this faith.

Where are the deserts, the third tropical region? Geographically they are all in roughly the same latitudes, lying on either side of the tropics of Cancer and Capricorn. They are in the high pressure region which divides the temperate world from the tropics, the belt which gives off the westerlies and the trades. They are in the horse latitudes, but they extend equatorwards as well. Most of the world's deserts are between 15° and 40° from the equator and in either hemisphere. In order of size the principal deserts are the Sahara (2,600,000 square miles), Australia (1,100,000 square miles), Turkestan (900,000 square miles), Arabia (480,000 square miles), Argentina (400,000 square miles), Colorado (195,000 square miles), Gobi (180,000 square miles), Kalahari (90,000 square miles), Thar (74,000 square miles) and Chile (74,000 square miles). As a size yardstick even the smallest of these deserts is larger than the British Isles, and the Sahara is thirty-five times larger than the Chilean desert.

It has been calculated that of the Earth's 200 million square miles of surface slightly over 29% is land. And of these fifty-eight million square miles of land at least 12½%, or over seven million square miles, is desert. If the co-called semi-desert areas are included the percentage goes up to 16½%, and the Earth's dry area to almost ten million square miles. It is a formidable total and would, if all joined together, form a square whose sides were over 3,000 miles long. One wonders how long mankind will be able to afford the luxury of possessing such unproductive blots on the map.

What is a desert? Dryness is, of course, the keynote, and heat the exacerbating factor. It is not dryness alone but the combination of much sunshine with little rain that creates the desert as well as the definitions of a desert. There have been various attempts at definition. For instance (using degrees Fahrenheit and inches of rain) a desert is said – by some – to exist if the average annual rainfall is less than a fifth of the average annual temperature. Assuming an average heat of 70° F (21° C), the place will be a desert if it receives less than fourteen inches of rain. Britain has an average temperature of 50° F (10° C) and, by this definition, would have to have less than ten inches a year to qualify as a desert. But there is more to climate than just heat and rainfall, and ten inches of rain in a place like Britain could never permit the place to become a Gobi desert.

Another definition, using the metric system, says a desert exists where the rainfall

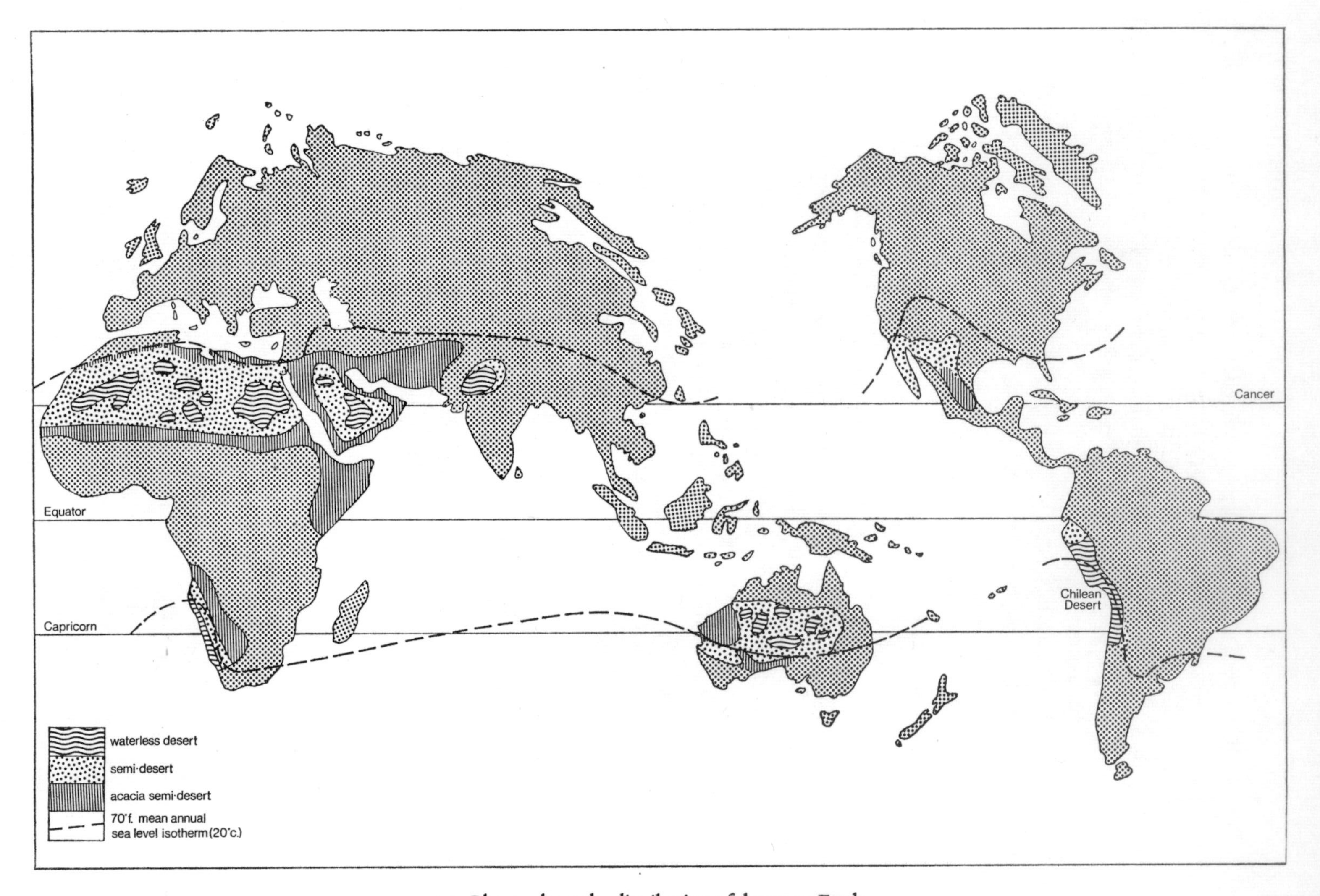

13. Plan to show the distribution of desert on Earth.

in centimetres is smaller than the average yearly temperature in degrees Centigrade plus 16·5. Assuming 70° F (21° C) and 35·5 centimetres of rain the temperature of 21° C+16·5 is a larger figure than the rainfall of 35·5 centimetres (14 ins); therefore the place must be a desert. Although the formulas are helpful a true understanding of dryness ought to take in other factors, such as humidity, wind, pressure and type of soil. The problem then becomes enormously complex. The reaction of various authors has been to say either that the place is a desert if less than twenty inches of rain falls in a hot place, or that anyone can see if it is a desert just by looking at it.

All reasoning based on average temperatures can so easily go wrong. Imagine twenty inches falling in some coastal area where the thermometer seldom moves far from 70° F (21° C). Imagine just as many inches falling within the depths of a large continent where the temperature can oscillate by 90° F (50° C) every day, as in the Sahara in summer. And does the rain fall when the ground is like a furnace or when it is freezing at night? The furnace gives little opportunity for seepage.

Thus the ratio of evaporation to rainfall is highly relevant. Put an upright bucket in an English garden, and the chances are it will have water in it at the end of a month. Only in the summer might the bucket be dry, despite fairly recent rain and a higher average rainfall. Only then might evaporation have exceeded this rainfall. At Ghardaia, in Algeria, the ratio of evaporation to rainfall is 59:1. In other words this Algerian town would have to receive over fifty-nine times as much rainfall as it actually receives for there to be any chance of the bucket's bottom being covered with water. In the depths of the Libyan desert, where it rains only every four or five years, the annual evaporation is 158 inches. In this arid case a bucket 13 ft 2 inches high and full of water at the start of a year would be bone dry by the end of it. Small wonder the place is a desert when rainfall has to compete with such a colossal ability to return water once again to the atmosphere.

Despite the obvious hostility of the desert environment, it is, generally, not without life. The plants which suddenly flourish, given half a chance and a drop of rain, have either been languishing as seeds or as bulbs, tubers or fleshy roots. The former, often called annuals (or, more appropriately, ephemerals because rainfall may not be a yearly event) have no particular adaptations to dryness save that their seeds can survive the drought. If there are two rains in the year each will bring out a crop of annuals, but each rainy time has its own crop. Those belonging to the other season can remain strangely dormant, despite the moisture. The foodstore kind of plant, with bulbs or tubers, grow their stems, leaves and flowers when they are able to do so after a rainfall.

Camel thorn belongs to a third kind of desert plant, the kind which can survive the dry times above ground. These plants often lie low, grow in places protected from winds, have few leaves or none, possess soft twigs containing chlorophyll, and have roots of colossal length which probe down in search of water. They can survive, but it is hardly surprising to learn that the desert has fewer species of plants than all the moister regions.

Instead it is surprising that the desert environment has about the same number of bird species and of mammals, and many more reptile species than in a nearby moist climate. Most of these desert animals are nocturnal, they live in holes, they are sand coloured, and have large feet. They are also well adapted internally to the hot, dry environment. They conserve water, they acquire water from their own metabolism, and they are more tolerant of heat and water lack. To be nocturnal is emphatically best because the night is cooler, calmer, moister, safer (from predators) and altogether a better time for finding food. (Humans should also dig a hole by day and travel by night if they are short of water in a desert.)

Dr J. P. Kirmiz has compared the desert jerboa's bodily characteristics with those of a white laboratory rat, with the one adapted to extreme dryness and the other no more adapted than the average human being. The jerboa (*Dipus aegyptius*), a rodent that hops in kangaroo fashion, can live for years without water on a diet of dry grains of corn. The rat can live for three days on such a diet. The jerboa's temperature is lower by 1·3° F (0·7° C). Its basal metabolism is lower by 68% and therefore the body does not heat itself up so much by its own chemistry. The animal reacts to extreme heat by becoming abjectly lethargic, and it can still exist when its temperature reaches 113° F (45° C). The rat's upper limit is 104° F (40° C). At this temperature of 104° F (40° C) the jerboa's heat production is 5·6 kilocalories per kilogram per hour while the rat's is 10·9 kcal./kg./hr. The rat is therefore double the jerboa in this respect, but its rate of heat loss is not double and the rat succumbs. Even without such detailed examination the rat is obviously needing and squandering more water. The jerboa eats dry food and produces dry faeces and concentrated urine, whereas the rat eats wet food and produces moist faeces and a lot of urine. The jerboa gains in every respect, and so survives in the desert. A human being has to produce over a pint of urine per day, and loses another pint in his exhaled breath. He also perspires. Liquid intake has to be similarly high, or he too succumbs. Man, in short, is a rat, not a jerboa, but he can do better than both if he looks after himself properly.

Sir David Brunt has outlined the probable tolerance levels in different conditions. There is risk of heatstroke to a clothed walking man at 70° F (21° C) if the air is totally saturated, but only at 110° F (43° C) if the air is totally dry. If he is sitting down both temperatures can go up by 10° F (5° C) or so. If he is both naked and sitting, there will be a heatstroke risk at 95° F (35° C) in saturated air, or at 160° F

(71° C) if the air is dry. A slight wind will push the lower figure up by a degree or two and the higher figure up to 200° F (90° C). At this extreme temperature the sweating rate is, or should be, 5·3 pints an hour. Naturally there would also have to be a similarly colossal intake of liquid to maintain the supply of sweat. If there was not, death would follow rapidly.

Even in more gentle situations, say 120° F (50° C) and 30% humidity (quite a feasible combination in the tropics), the sweating rate for a naked man resting in still air is almost a pint an hour. Mankind is therefore able to cope with temperatures considerably higher than those which would kill the jerboa or the white rat, but only with the most lavish use of liquid. He might as well be sitting under a shower as pouring out pints of sweat every hour. *Dipus aegyptius,* living on seeds and other arid bits of vegetation, cannot afford such luxurious indulgence.

Even the hottest deserts have relatively cold nights and cooler winters. So even here the seasons and the cycles are firmly in evidence, and firmly affecting all living things in the area. Unlike the temperate world, the heat of the desert is more savage than the cold and a greater hazard. However the imposition of a cyclical rhythm is similar, and so are its effects. Plant and animal life must withstand a variety of circumstances, and possess a good measure of variable adaptation. The law is the same, pole or tropic, cold wet Europe or hot dry desert; and it is just as much so in the deluge-to-drought world of the annual monsoon. This phenomenon is a remarkable extreme of the tropical propensity for a rainy season. More like the downpour of some river than mere rain, it deserves fuller study.

The sub-continent of India is admirably suited to this kind of weather. As Dr A. Austin Miller wrote (in his book *Climatology*), India lies 'between the greatest landmass and the warmest sea, shut in by the loftiest mountains, backed by the highest plateaux of the world, and provides the best conditions for a monsoon'. The Himalayas, the Sulaimans and the Burmese mountains are effective insulation against the climatological happenings of the rest of the Asian continent. If some future agency could remove these mountains, the effect on India would be to provide a more regular rainfall and to open up some of the desert country to the north of the mountain range, country at present frustrated by the inability of water-bearing airstreams to reach it.

In winter India and Pakistan are high pressure areas, being part of the enormous Asian high pressure system; during that period the sub-continent is then a dispeller of dry air. But in the Indian summer, when air has to rush in to fill its northern low pressure, an area already cooked so fiercely by the sun, the air arrives inevitably from the south, inevitably loaded with warm, moist, Indian ocean air, and it is promptly relieved of its load over the land. India then receives the burst of the monsoons, and relishes it.

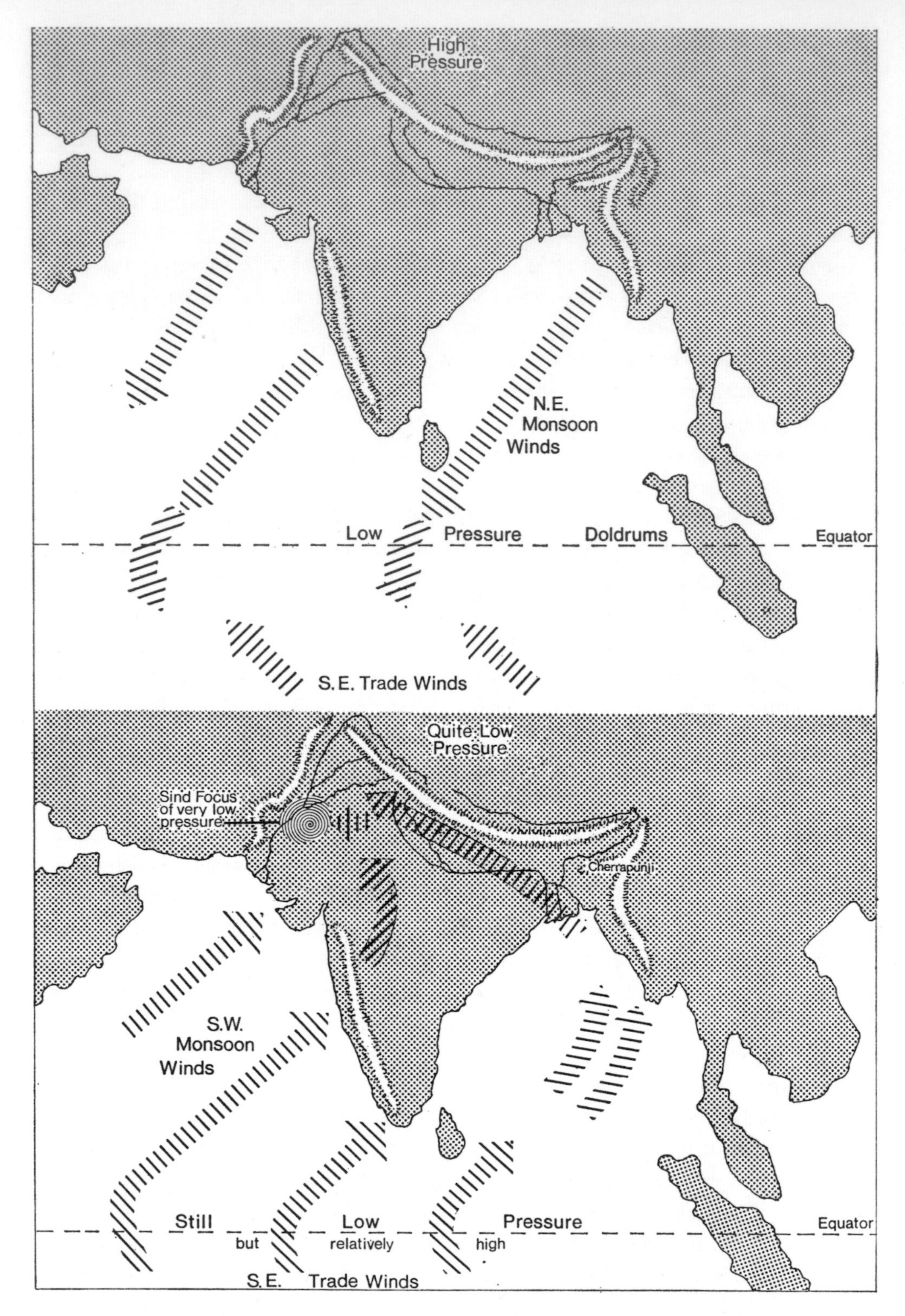

14. The top diagram shows how, during India's winter, high pressure builds up in Asia and pushes air outwards to meet the south-eastern Trade Winds near the Equator. This clash produces the static area known as the doldrums. The season changes when an even lower pressure builds up in northern India. The direction of the winds is now reversed so that the moisture laden winds from the ocean blow over a dry land-mass, causing torrential rain – monsoon conditions.

Essentially, therefore, the monsoonal situation is one characterized by a shift in wind direction. Other places than India have monsoon climates, but nowhere is the situation so marked as in India. Even so the changes which actually precede the burst, and which cause the life-giving deluge, are really quite small. They act, in many ways, like the last few cracks in the dam that suddenly cause the whole structure to give way and flood the land.

In May much of India and Pakistan is languishing under drought. The hot season is getting hotter. The southern tip of the continent is getting some of the benefit of its closer proximity to the more regular equatorial climate, and the Ganges area is getting some rain, but almost the entire continent is panting for the summer monsoons. Bombay has had no rain to speak of since the last monsoon ended. There is a small pressure difference of 0·2 inches between Sind and Ceylon, and there is also a slight wind blowing from the Indian Ocean towards the sub-continent, but these gentle windstreams tend to cancel each other out. The situation seems stable, and the few rainstorms that do occur are entirely the result of local conditions.

Suddenly, and for no violent reason, the situation dramatically alters. The south-east trades overcome the local doldrum-like conditions in the Indian Ocean. They then advance into the Arabian Sea and are soon deflected straight for India. The pressure drop from Sind to Ceylon is then greater, the continental effects (expelling hot air) of India are quickly overcome, and warm air, laden with water after some 4,000 miles of travel over the world's hottest ocean, advances upon India at great speed. The dam is breaking. The deluge is due to begin.

It usually does so appropriately as a storm. The English may welcome the swallow as a sign of better things to come, but the all-important date for India is the average date of the monsoon burst, 5 June at Bombay, 10 June in the central provinces, 15 June at Bengal, 20 June in the eastern provinces, and the end of the month for Delhi. The welcome wet tide sweeps over the country from the west towards the north and east. Bombay's long drought – an average of 1·3 inches of rain since the end of October in the previous year – leads to twenty inches in June, twenty-four inches in July, 14·5 inches in August, and 10·6 inches in September. But that is all, or nearly so. It started abruptly, and it abruptly ends, for October has an average of only two inches and November has none. The four wet months have poured a total of some seventy inches of rain on the land; the next eight months will produce just 70 mms. If the monsoon tap is switched off too soon the rice crops can fail from lack of growing time. If it is not switched on soon enough the drought's prolongation may also be disastrous. Either way the monsoons are omnipotent.

The situation is yet more imbalanced whenever the inrush of hot wet air has to ascend to any altitude. A hundred miles to the south of Bombay is Mahabaleshwar at a height of 4,540 ft. From June (after the burst) to September a rainy day means, on average, two and a half inches of rain (or twenty-five times the ordinary amount of a wet English day). And in that four-month period there are 116 rainy days out of the possible 122. July is peak month with a total of 100 inches (or over three inches a day) and 250 inches fall in the monsoon period, in this brief, wet, third of the year.

The arrival of the Indian monsoon in high summer, and at the peak of the summer solstice, has the effect of lowering a temperature which might otherwise be expected to rise in customary summer fashion to reach a maximum in July or August. Bombay is 75° F (24° C) in January and February. The sun starts climbing higher, to make the average temperature 78° F (26° C) in March, 82° F (28° C) in April and 85·8° F (30° C) in May. Then comes the doubly welcome rain. Not only is it wet, but the temperature falls to 83° F (28° C) in June, and to 81° F (27° C) in July and August, when it could so easily be a whole lot worse. By the time the rain stops, the sun is returning once again to the southern hemisphere, and the temperature falls to the wintry minimum of 75° F (24° C). Unfortunately, although the rain cools the time of maximum heat, it also arrives at the period of maximum evaporation, of maximum run-off, and when so much of the water will be gone before anyone has a chance of using it. In wintertime it would disappear less quickly from the soil, but there would be a high price to pay: the long summers would then be both extremely hot and desperately dry.

So much for the hot places. Either hot with rain all the year round, or hot with two dry periods, or hot with one dry time, they are a world of their own compared with the cool- or even the warm-temperate regions. But there is still a need for adaptation to a changing situation, to night and day, and to the annual cycle, so predominantly coloured by rain or no rain. A lack of rain stops growth just as emphatically as a lack of warmth or a lack of light. Warmth and light are generally available in the tropical world; rain not always so. Adaptation, and the need to respond to a changing environment, is still of paramount importance.

11 Cataclysms

The regularity of the astronomical movements which leads to the smooth running of the seasons implies a regular succession of fairly predictable events. On the other hand the fact that there are mountain ranges, and land masses and ocean masses, and the Earth's crust is still cracking, and the planet itself is losing heat, and each hot day is followed – probably – by a cold night, and that the sun's radiation lands on Earth unevenly – all these and many other factors indicate a likely unevenness in any assumed regularity. Sure enough there are typhoons, hurricanes, waterspouts, earthquakes, tidal waves, droughts, deluges and eruptions. Krakatao did blow up. The new volcano, Surtsey, did appear out of the sea off Iceland. Seventy-two inches of rain have fallen in an Indian day. There are a million earth tremors every year. Meteorites have landed. Avalanches have buried villages. Those stories of grapefruit-sized hailstones are true. (Frank W. Lane has collected countless stories of all kinds of disasters in his excellent book *The Elements Rage*. All students of natural disaster should read it.)

Some extra-terrestrial scientist, having been informed in general terms of the planet Earth's properties, of its age and composition, and of its position with regard to the sun, might reach an opinion about the nature of our seasons, but he would still be baffled about the likelihood of any exceptional circumstances. He would appreciate the existence and probable direction of winds but be entirely confused about their potential power and destructive capability. How often would hurricanes happen? How damaging would earthquakes be? And would the consequent tidal waves be high or severe enough to flood low-lying countries? He could argue that, as life exists on Earth, the situation cannot be too bad. On the other hand life could exist if the cataclysmic events were a lot worse than they actually are. Only a visit to Earth would enable this observer to discover the true extent of their damage and frequency.

Most cataclysms are seasonal, and many are likely to occur at a particular time of day. These more predictable disasters have the sun as one of their primary causes, and include high winds, hail and lightning, extreme rains and floods. Far less predictable are those caused by the Earth itself, the tremors and the eruptions, although these occur predominantly in certain areas. Fortunately the most random events, the landing of meteorites on the Earth's surface, are rarest of all. Life would still be possible if our atmosphere was less protective against meteors, or if there were more of them with either the mass or the speed to penetrate to ground level, but it would be an unnerving existence for all conscious forms of life on Earth if these bolts from the blue landed and devastated some area, say, once a day. (Robert Benchley is alleged to have given the fear of meteorites as an explanation for his unwillingness to sit in deck chairs.)

Rarest of all cataclysms are explosions like Krakatao. That island blew up in 1883, when four cubic miles of rock were hurled into the stratosphere, and barographs all over the world recorded the pressure changes. Exactly a year after that big bang one spider and a few blades of grass were found on the island. Twenty-five years later 202 non-microscopic species of animals were found in a three-day search. Eleven years later a further expedition found 621 species in sixteen days, and half a century after the eruption 880 species were found in a two-week period. Life can be destroyed in an area, but will soon make good the damage.Even Bikini and Eniwetok, pulverized and radiated by fifty-nine nuclear blasts between 1946 and 1958, were back to a kind of normality by 1964. Destruction of the topsoil meant regeneration was slow, but life was living there.

Natural devastations are impressive and the loss of life can be tremendous but, bearing in mind the 200 million square miles of the Earth's surface, and the one-third which is land, most places are quiet most of the time. Newspapers somehow emphasize this point by making such dramatic play of fairly minor disturbances, particularly if they are published in areas not subject to major disturbances. Hurricanes are said to be sweeping the country, and maybe they cause a couple of people to be killed by falling tiles. Blizzards are reported when only a few sheep have their lives put in jeopardy. The parts of the world which do customarily have hurricanes and blizzards must find such headlines comical because, when the atmosphere does pull a few tricks out of its sleeve in the vulnerable areas, the effects can be catastrophic. Whole houses are thrown about with their tiles intact and few people in such zones have time to worry about sheep.

A hurricane, or typhoon or cyclone (Asia), or baguio (Philippines), or willy-willy (Australia), is the most destructive kind of wind. One storm can cover half a million square miles and last three weeks. These winds form above a warm sea, probably when its temperature is 80° F (27° C) or more, and they need the effect of the Earth's spinning to set them revolving correctly. Hurricanes cannot form at the equator and they do not necessarily form even when the situation may seem right for them.

On average there are eight Atlantic hurricanes a year, but in 1914 there were none. They occur in late summer and autumn when the water is warmest, they always revolve anticlockwise in the northern hemisphere – as do all low pressure area winds – and they usually travel west or north-west. In the southern hemisphere everything is reversed. There they arise six months later – in the southern late summer and autumn – they revolve clockwise, and they usually travel east or south-east. The pressure drop involved may be a couple of barometric inches or 140 lbs load off every square foot of surface involved. The top windspeeds may be 250 m.p.h. The calm eye in the centre, usually about fourteen miles across, can give a false assumption that the storm is over, but this is only a misleading interval before the winds roar with similar voraciousness from the opposite direction.

The most destructive hurricane in American history was Betsy which came ashore in September 1965 to travel from southern Florida to Louisiana and cause damage worth $1,500,000,000 *en route*. Camille of 1969 was almost as bad. The most deadly was in 1900 (before girls' names were adopted) when 6,000 people died in Galveston, Texas, mainly as a result of the water dumped on the town by the winds. India, being so crowded, has more to lose in human terms: 300,000 died from a typhoon's wave on 7 October 1737. In 1942 an observer in Bengal said he saw the sea retreat for twelve miles and then return as a wall of water thirty feet high. Anyone who has seen the sea retreat once, or has read of this kind of occurrence, should not wait to watch its return, however spectacular this might be. He should be making for high ground without delay.

As with hurricanes, the effects of tornadoes and water-spouts can be dramatic and devastating. They cover far less area and last for less time, but they make up for these deficiencies by being far more violent during their – probably – minutes of life, as against the days or weeks of a hurricane. Tornadoes are generally summertime events, occurring from March to October in the northern hemisphere, but they can also break this rule. (One did in January 1969 when it killed twenty-eight people in central Mississippi.) The United States has the world's worst tornadoes, and also more of them than anywhere else. The most frequent place and time is in the middle south of the continent from April to June. Traditionally they are afternoon or evening events which, unlike hurricanes, usually travel from south-west to north-east. Destruction can be tremendous in the path they decide to take, but the average length of their path is only two miles, and the average width of each tornado's funnel is only 1,000 ft. Living things in these average 242 acres of path can suffer tremendously, and a tornado often kills more people than a hurricane, although the destroyed area is relatively small. Kansas is hit more frequently by tornadoes than anywhere else in the world receiving, on average, twenty-four twisting visits a year. Even so, assuming their destructive existence to be average, this means that only nine square miles of land are ravaged by tornadoes each year.

Waterspouts, the aquatic equivalent of tornadoes, are naturally less well recorded and less damaging. They are also summertime events, occurring mainly by day in warm seas – although, of course, night-time sightings are likely to be rare. They are perfectly capable of sinking large ships, and have frequently done so.

It is remarkable how these destructive terrestrial phenomena are so conveniently rare. Although they are natural phenomena and have – presumably – been occurring since the beginning, it is hard to imagine any form of life adapting itself to one or other of them. A palm tree blown over almost horizontally, but surviving a hurricane, cannot really be said to have adapted itself to hurricanes. If the tree survives, its survival is of course advantageous, and therefore all the arguments in favour of natural selection can apply, but these natural cataclysms are of quite a different order to the ordinary hazards of life. A tornado can pull trees up by the roots and throw them down somewhere else. A hurricane can flood a desert. An avalanche can swamp a valley. In such circumstances the customary struggle for survival becomes a lost battle almost the moment it begins. So too with hail, lightning, earthquakes and volcanoes. The greater and more important struggle is to invade the devasted area, to recolonize, to see that Krakatao lives again.

Most places know hail, but only the warm-temperate or sub-tropical land masses suffer the really damaging storms. In financial terms hailstorms cause less damage in the US than hurricanes but more than tornadoes. Hailstorms can only occur under thunderclouds. The big stones, perhaps three inches in diameter, are likely to come only from very big clouds. The onion appearance of a large hailstone, with its many concentric layers, suggests that such a stone becomes big only by a succession of ascents and descents, until such time as the updraught is no longer capable of supporting its load of ice-cubes. This may frequently be the case, but recent radar work has shown that such journeys are not crucial; one powerful updraught on its own can lead to the build-up of a big stone. (How curious that radar is managing to solve so many entirely different problems bordering on the seasons such as the migration of birds, the rotation of other planets and the creation of hailstones, all discussed in this book.) Recent work has also shown an association between jet-streams, the high-altitude high-speed winds only discovered during World War II, and hail-stone formation. As jet-streams over the United States tend to travel from north-west to south-east or from south-west to north-east the hail damage on the ground follows similar lines. Hailstones can be damaging not only in their initial pulverizing onslaught but in the way they cover the ground temporarily with ice.

Lightning is often the most impressive natural phenomenon, but the damage done by these discharges is small. Thirty million horse-power are sometimes consumed just by the production of thunder, and the energy involved in the lightning bolt itself may be 400 times greater, but these fantastic energies are usually dissipated harmlessly. The greatest disasters have occurred when lightning has sparked off something else, such as a store of gunpowder. At any single moment in time about a couple of thousand thunderstorms are occurring, most of them in the damp tropics.

There are plenty of theories, but no general agreement about the cause of lightning. Large animals suffer more than men during storms possibly because their four legs cover more ground. Vegetation suffers most because lightning can start forest fires. Admittedly forest fires are frequently started by man but many – perhaps 50% – of California's forest fires are thought to be caused by lightning. Forest fires are also frequent in areas where man is not so thick on the ground, such as central Brazil.

Finally, there are floods which cause immense damage by submerging the land beneath fresh water or, worse still, beneath salt water. They may be frequent – many places are flooded annually – or irregular. In their surprise, irregular floods can be more damaging for vegetation is often quite unsuited to such an inundation, particularly if the water is salt water. It is hard, of course, to generalize about the location of unpredictable floods. They can occur in deserts or in monsoon regions or whenever the normal drainage system goes wrong. Seasonally they can only occur at the rainy time of the year, normally when rivers are already high and the ground already saturated. The melting of the snows in spring is a dangerous time, and wind and water together can cause flooding, as with the 1953 disaster in East Anglia and Holland. A high tide plus a gale plus a drop in atmospheric pressure then sent the North Sea surging over the dykes. The Mississippi has had some appalling floods, but India, Pakistan and China overshadow the death-roll in the United States by a wide margin.

While floods and fires are devastating they can have extremely beneficial consequences. In their inundation they cover the old land with new minerals, and thereby restore its life. In a sense this is the story of life itself, for death makes way for life, and destruction enables more life to be created. The ravaging of a forest fire permits new growth to establish itself in a soil then much enriched by phosphates from the fire. The plant which can establish itself first after some destruction, and be the first to rise from the flames, is proving itself able to take advantage of disaster.

The blackened face of Krakatao took longer to recolonize than a similarly devastated mainland would have done because Krakatao is an island; but, even so, scientists were amazed at the speed with which life returned there. Presumably most of the recolonization came from Sumatra and Java, fifteen and twenty-five miles away. By 1933, when the expedition arrived after half a century to find 880 species, it found the island remarkably similar in its inhabitants to nearby islands which had not suffered. There was even a small, but inevitably young, forest covering much of the island. There were thirty species of birds and 700 species of insects. Of course birds and many insects can fly, and therefore the barrier of seawater separating Krakatao from elsewhere would be less of an obstacle to them; but four species of mammals had arrived and six kinds of reptiles, including a python. Of spiders and molluscs there were 92% as many species as on comparable nearby islands. It will not take centuries or millennia for the island to achieve its former stock, as many had believed. Even a single century may be sufficient.

The seasonal excesses – the extremely cold winters or the drier-than-average summers – ensure that the organisms able to survive such abnormality benefit accordingly. The cataclysms, the highly irregular changes in the environment, are also able to bring benefit – to some. It is an ill wind, as the old aphorism puts it, that blows no good. Rose-bay willow-herb is a case in point. It is even known as fire-weed due to its enthusiasm for growing on scorched earth; and the bombed areas of London sprouted this plant most energetically in the last war. Charles Elton quotes (in *Animal Ecology)* the case of the shipworm which benefited from dry summers. A lack of rain caused the inland freshwater to become saltier. The saltiness, in turn, enabled *Teredo navalis* (the shipworm) to survive in abnormal areas. Ship hulks suffered more from the extended burrowing of this wood-eater in 1730, 1770, 1827 and 1858, for the dry summers had brought ship repairers good business.

Some ill winds, such as cyclones and hurricanes, would appear to do nothing any good, in that the destruction to life is so complete; but the subsequent recolonization must end each time with a slightly different proportion of organisms. Krakatao will probably not be quite the same again. Cataclysms play their irregular part in the relentless re-ordering of creation. A volcanic eruption, one imagines, is the ultimate disaster, as molten lava pours over the soil. In fact, as is demonstrated by the willingness of people to live on volcanoes, mixed lava and volcanic ash are broken down by the weather to form exceptionally good soil. The ash falls short of being the complete natural fertilizer only through its lack of nitrogen, and so even this ultimate in ill winds blows much good. In conclusion cataclysms are rare, they are beyond the normal range of environmental parameters, but they play a part. They devastate the land, and the land then lives again.

12 Life Without Seasons

However equable the climate anywhere, however regular its days and nights, life in that place is still subject to annual and daily rhythms. Surface life has always responded to these rhythms, and its own cycles follow suit in their own particular fashion. But what about a world without seasons, a place totally different from the tropical, temperate and polar zones? It is because they are seasonless that caves offer an environment worthy of study. After all, only a very few yards and twists from the entrance of a cave is a dark world, unaware of the passage of night and day, and largely unaware of seasonal change, of warmth and cold, of rain and drought. Instead, it has its own damp, cold, thick-walled and changeless climate. Therefore animals living perpetually in such a place, independent of outside influences, might be expected to live differently, to be without a breeding season and without a diurnal routine.

First, the environment of a cave. Go into such a place on a warm day, even without measuring instruments, and very soon there is a marked change in light intensity, temperature and humidity. Before very long it is probably dark, very cold and the air is water saturated. Naturally the extent of these parameters varies from cave to cave, particularly if the cave has two entrances or more – thus permitting some kind of through draught. In an ordinary one-hole cave the internal slope is also important. If it is inclined upwards, and if the cold air can therefore slide out of the entrance, some circulation of air will be possible. If inclined downwards, and if there is only a single entrance, this kind of cave will be the most static.

Measurements have been taken, for example, in the stalactite cave of Baradla in Hungary, which has such a downward slope 150 ft long. Maximum and minimum temperatures at the doorway during one whole year varied from 35° F (1·7° C) to 63° F (17·2° C). A mere ten steps from the entrance the annual extremes had shrunk to 44° F (6·7° C) and 53° F (11·7° C). At the far end of this short tunnel, despite being only 150 feet from the doorway which experienced a temperature range of 28° F (15·5° C), the range was only 3° F (1·6° C); the maximum and minimum recorded were 51° F (10·5° C) and 48° F (8·9° C). During this time the greatest relative humidity was everywhere 100%, but during the drier times there was considerable difference in the minimum air humidity. At the doorway, which was on the southern slope of a hillside, the air could fall to a humidity of only 19%. Ten steps within the cave it never fell below 77% and at the end of this cave – which had no stream running through it – humidity never fell below 96%. In other words the end of this short shaft only varied its temperature by 3° F (1·6° C) throughout the year, and its humidity by 4%. Few other climates in the world are quite so consistent.

The change in light intensity can be equally remarkable. For example, the Hölloch cave in the Allgäu of the European Alps starts off with a vertical shaft about 26 ft wide and 230 ft deep before the ramifications of the cave itself are reached. Even in this shaft, which points straight towards the heavens, light intensity falls away sharply: at 16 ft it is 90% of the surface value, at 32 ft 30%, at 50ft 10%, and at 65 ft only 4%. Therefore light becomes virtually non-existent even at the foot of such a vertical shaft, let alone a horizontal passageway as at Baradla. To be perpetually dark as well as climatically consistent makes the environment even more abnormal. So what animals and plants do live in caves, and how do they make out?

To a large extent cave dwellers have tended to make use of the place as a refuge, either for the night, or for the day – as with bats – or for the entire winter – as with some bears. The great period for refuge of this kind was during the recent glaciations: the savage drop in temperature sent mammals of all kinds into these shelters. Man was on this list, being a considerable cave-dweller until neolithic times. Even the mammoths took refuge, so much so that the Russian peasants who unearthed their huge skeletons from time to time imagined they had been giant burrowers and named them after the *mammot,* the Russian word for mole.

The huge cave-bear, *Ursus spelaeus,* was far more of a permanent cave-dweller but, oddly, very rarely drawn by early man, although such cavernicolous cohabitation must have been of enormous consequence to both. Cave hyenas lived in most European caves, and could obviously have done well without ever leaving the cave entrance because countless skeletons of cold tundra fauna dating from the late Pleistocene have been found in the caves. Whether they were dragged into the cave by man or beast or merely died there is unknown, but the species found so far include wolf, otter, rhinoceros, Irish elk, reindeer, auroch, bison, horse, lion, beaver, lynx, fox, pig, musk-ox, ptarmigan, mole, badger, hamster, marmot, ass and chamois, as well as the mammoth, hyena and many bears other than the cave-bear.

Whether most of these animals entered caves of their own accord or not, they are not generally considered as cave species today. The most cave-loving of them all, the cave-bear, became extinct like so many of the largest forms at that time, and bears today make their winter dens in a multitude of places. However there

Volcanoes Natural cataclysms may be caused originally by the sun or, in the case of earthquakes and volcanoes, by movements within the earth. These are relatively rare but hazardous because of their unpredictability. Volcanoes can occur on land or in the sea, chiefly in areas of weakness in the earth's crust. Rarely regular, volcanic activity can sometimes be predicted by changes in the magnetic field of the area or by slight subsidence prior to an eruption.

LEFT: Volcanic activity can create totally new islands. The Surtsey volcano in the Westmann Islands off Iceland was thrown up above sea-level in November 1963. On the eighth day of eruption (shown here) the new island measured seventy metres high.
ABOVE: Lava flowing into the sea from a volcanic crater.
BELOW: The largest volcanic eruption in recent times occurred in 1883 at Krakatao, Java: four cubic miles of rock were thrown into the stratosphere and all life on the island was apparently destroyed. A year later, however, a spider and a blade of grass were found, and twenty-five years later over two hundred animal species were counted. Fifty years after the explosion the island was again covered with young forest and supported a plant and animal life similar to that of the neighbouring islands.

3

Earthquakes Earthquakes are produced by movements under the earth's crust and range from the slightest tremor to the complete devastation of hundreds of thousands of square miles. Two main belts of the earth, where mountain-building processes are still going on, are particularly prone to earthquakes: one stretching from North Africa to the Balkans, the other covering Japan and part of the Pacific.

LEFT: The results of the earthquake in north-west Japan in June 1964; as many as thirty per cent of the buildings were completely destroyed in some areas.

ABOVE: The same hotel before and after the Agadir earthquake of February 1960; over 10,000 people lost their lives.

Storms A hurricane or typhoon can cover half a million miles and continue for three weeks. It is a late summer or autumn phenomenon, and has a deceptive 'calm eye' in the centre.
LEFT: The Tiros weather satellite records the birth of a typhoon.
Tornadoes (BOTTOM RIGHT) and their aquatic equivalent, waterspouts (BOTTOM LEFT), are more violent than typhoons but cover far less ground and usually last for a matter of minutes. The United States is the country most prone to tornadoes: Kansas has an average of twenty-four a year.

Tidal waves are a result of submarine earth movement, and can be hundreds of feet high.
LEFT: A hillside torn away at Lituya Bay, Alaska, by the highest tidal wave of recent times on 9 July 1958.
BELOW: A hurricane in Florida. There are roughly eight Atlantic hurricanes a year, with windspeeds of up to 250 m.p.h. In 1900 six thousand people in Gaveston, Texas, lost their lives in a hurricane; in September 1965 the hurricane 'Betsy' devastated an area from southern Florida to Louisiana, causing $1,500 million worth of damage.

Floods Floods are usually caused by rivers overflowing due to high rainfall, melting snow or ice, high tides, high winds, earthquakes or any combination of these.
ABOVE: Water from a flooding river sweeps down the main street of Putnam, Conn., U.S.A., in 1955.

TOP LEFT: The flooding of the Arno in November 1966 devastated Florence with the worst floods ever known in the area. Thirty-nine people were drowned; twelve square miles were submerged; over a thousand works of art were damaged.

BOTTOM LEFT: Floods can reach extremely high levels, as shown by this cow left stranded in the tree-tops at Nashville, Tennessee, after severe flooding.

Avalanches and dust storms LEFT: Avalanches may consist of powder snow blown into the valley like a cloud, but the more dangerous type, caused by melting snow and ice in spring, carry rocks, earth and trees with them. They may be caused prematurely by a warm dry wind; if conditions are right a sudden noise can set an avalanche rolling.

ABOVE: A dust storm near Majrag, Jordan. Although dust is present in all air, dust storms occur only in deserts or hot dry areas where topsoil particles are extremely light. Thousands of tons of dust may devastate huge areas, endangering people, animals and machinery.

Fire A forest fire can be caused by lightning or by a mere cigarette stub.
ABOVE: Bush fire in Western Australia.
RIGHT: The Empire State Building is struck by lightning. Energies amounting to hundreds of millions of horsepower can be dissipated in a flash of lightning, but disasters are rare. At any one moment roughly two thousand thunderstorms are taking place throughout the world, mostly in the tropics.

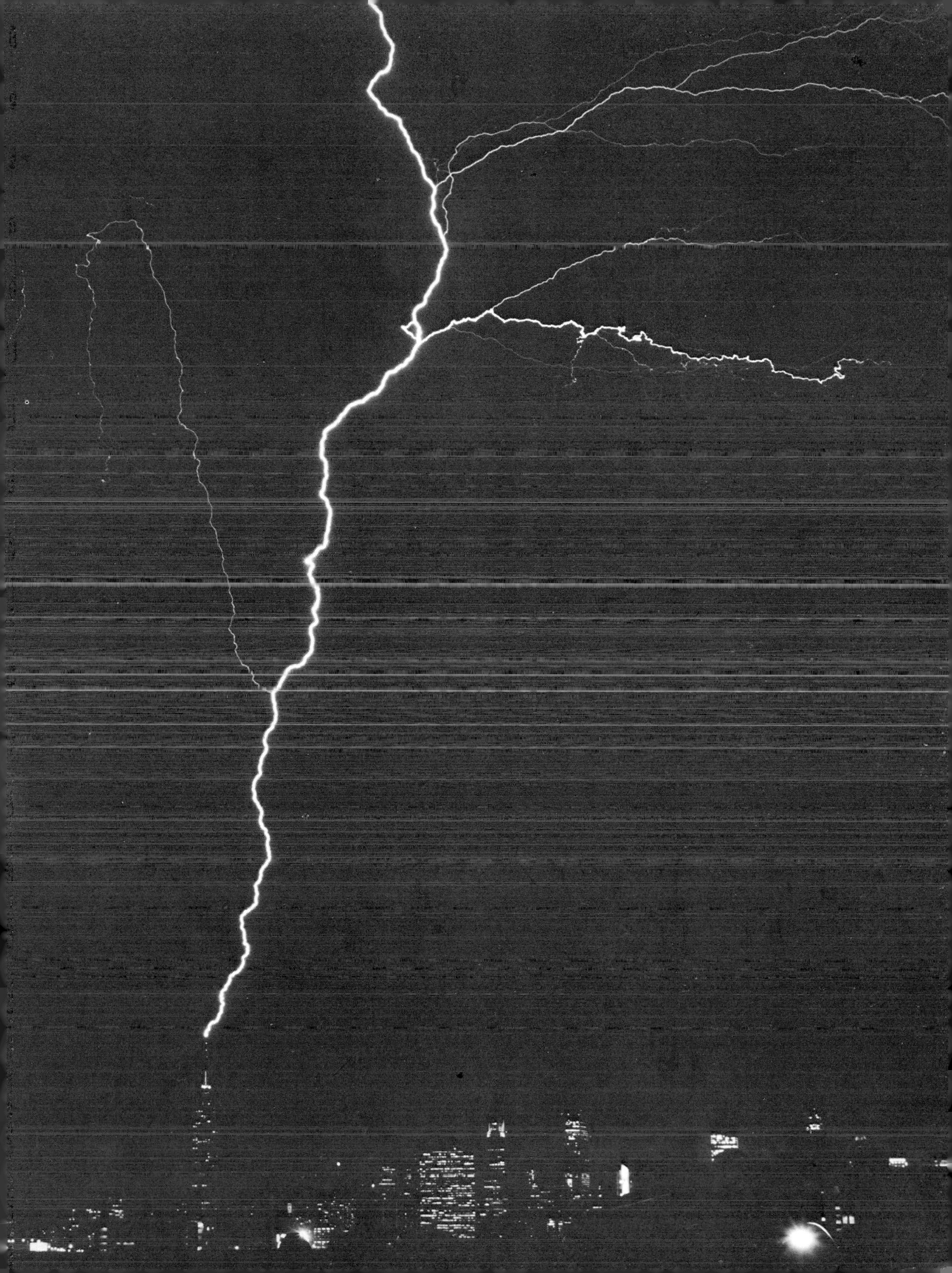

Europe's most famous volcano, Vesuvius, has remained active since the great eruption of AD 79 which destroyed Pompeii and Herculaneum. Recent eruptions have occurred in 1873 (shown above), 1906, 1929 and 1944. Lava flows have made the mountain slopes into an extremely fertile wine-producing area.

Subagan on the island of Bali, where two hundred people died in the Agung eruption of March 1963. Suffocating fumes are as much the cause of death as boulders and molten rock.

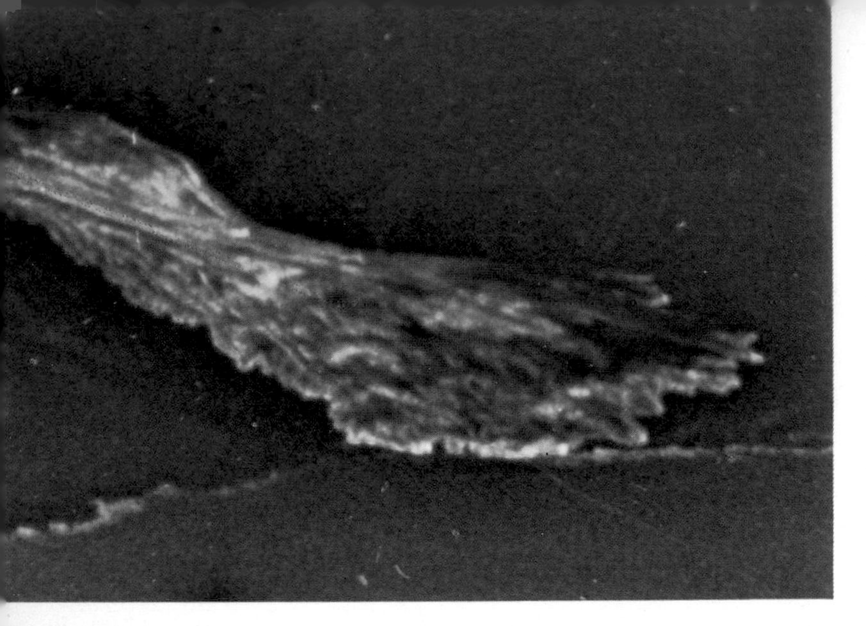

The renewal of life Cataclysms can play a creative as well as a destructive part in the cycle of creation.
LEFT: Molten lava when cooled and broken down by the weather forms exceptionally good soil.
BELOW: The small plant Sword Fern growing in lava after the Kilaues Iki eruption of 1959. Life reasserts itself quickly: even the islands of Bikini and Eniwetok, pulverised and radiated by fifty-nine atomic blasts over a period of twelve years, were on the way back to normal by 1964.

is still a group of mammals that make considerable use of caves. In fact one wonders if they would survive as well without these freakish geological caverns, for bats all over the world live in caves, while just a few species, such as *Pteropus medius,* the fruit bat, roost in trees. The largest and longest cave in the world, the Carlsbad Cavern, was only discovered after a black cloud of millions of bats had been seen emerging from the ground. Cave swallows also emerge in large numbers, the so-called salangane whose nests make bird's nest soup. They breed mainly in caves, and so do the guacharo birds of South America; but, however cave-loving, they and the bats are still visitors. It is the permanent residents that ought to be more revealing about their strange environment.

For example, there are genuinely cavernicolous insects. They are frequently blind and pale, without wings and with long antennae. An even greater modification is the atrophy both of their tracheal network of respiratory tubes and of the stigmata which lead to them. Presumably, owing to the permanently moist atmosphere, such creatures breathe through their skin. These insects will die if brought out into the external world because their moist oxygen-collecting bodies soon dry up.

Spiders are also found in caves and are often white, with bodies made large by considerable growth of their sensitive hairs. (Anton Lübke, of Germany, refers in *The World of Caves* to a spider the size of a human hand living in Pyrenean caves.) In the water there are many amphipods and copepods, minute crustacean forms, which are again frequently blind and white. In fact most true cave forms are crustaceans of one sort or another, and blindness and whiteness are the general rule. The several forms of cave snail are an exception, both in being molluscans and in retaining pigment and eyes.

This pattern of lack of pigment and of vision can be superficially explained by saying there is no need for either of them, but such a slick answer has recently been causing disquiet. For example, it has been suggested that the prevailing lack of oxygen metabolism has led to the lack of pigmentation. It is argued that the creation of pigment follows the oxidation of certain organic compounds, and if there is a general lessening of oxygen availability there will be a lessening of pigmentation, particularly as it has no selective advantage in a black cave. Vision is also dependent upon certain pigments, and these too might be the victims of a lack of oxygen rather than a lack of light itself.

That all cavernicolous forms consume less oxygen is undeniable. Two amphipod species were once examined, the surface form being similar to the cave form in most respects. It was found that in respect of oxygen consumption the cave species used between a quarter and a fifth as much as the surface amphipod. Two more similar shrimps were used in another experiment. On this occasion the cave and surface forms had their survival times measured, having each been given a sealed container to live in, and with both containers having an identical quantity of oxygen in the water at the outset. The surface shrimp survived for 272 minutes, the cave shrimp for 892 minutes. Animals existing in this cold world saturated with moisture are used to acquiring very little oxygen, and poor respiration must be responsible for much of the lethargic behaviour expressed in caves.

The constancy of temperature has also made some cave fauna peculiarly susceptible to temperature change. *Prostoma clepsinoides,* for example, is a worm with a remarkable lack of tolerance. It will die if the water is either cooled to 41° F (5° C) or warmed to 59° F (15° C). Many creatures prefer a narrow temperature range, but will not die if temporarily outside the preferred narrow margin. Prostoma's limits of tolerance are somewhat similar to mankind's warm-blooded internal range (say 91° F (33° C) to 109° F (43° C)), but an internal milieu cannot really be compared with an external environment. Obviously such a creature has not been subjected to seasonal vagaries for a long time. Equally obviously, for such is the underlying theme of this book, the lack of seasonal experience makes the animal totally unsuited to a changing world. Its powers of adaptation have been muted. Instead it is suited to constancy, not to variety, and it perishes at the first breath of change.

A misfit for all generalizations about cave fauna is Proteus, the cavernicolous egg-laying amphibian found principally in Yugoslavian caves. Despite being a traditional white in the darkness in which it lives, the males become grey in the breeding season and dark spots appear on their tails. The females, normally white like the males and slightly tinged with pink from their blood, become redder in the breeding season. Yes, there is a breeding season even in their nether world: it is May. Why and how there should be a breeding season is unknown, even though Proteus has been an object of wonder and study since its discovery nearly two centuries ago.

Perhaps its other curiosities have been too absorbing. Instead of an atrophied respiratory system, as with so many cave forms, the Proteus mechanism is duplicated. There are both internal lungs and three pairs of external gills. At least it is normal for a cave form in having poor vision, for its pinhead eyes are covered by skin. Nevertheless, true to its abnormality, Proteus is sensitive to light. Experiments have proved that it shrinks away from heatless light, and people have assumed that sensory organs on its skin are acting as receptors. Finally, when Proteus is brought into the light, it darkens. It is thought that this pigmentation is merely the result of greater oxidation, and therefore a feature of the chemistry involved in pigment production (already referred to) rather than any adaptive system of camouflage retained over the centuries. In this one area Proteus seems to be entirely normal.

Very little work has been done on cave breeding seasons. Most cave-dwelling species can be observed at all stages of their development, whatever the season. This is certainly true of the well-known fungus gnat, *Polylepta eptogaster,* which can only live in caves and quickly dries up if brought outside. It is also true of *Plesiocrerius lusicus,* the spider of German caves which never emerges and always breeds independently of the outside seasons. Cave forms also disregard the outside world in possessing many slightly different forms of the same species, and many different species of the same general type. It would therefore seem as if mutations are either more common in a cave (it is hard thinking of reasons to explain this possibility) or are more likely to survive in the population.

The cave environment is so odd in so many ways that direct comparison between forms of cave life and surface life are very difficult to make. Caves are not just worlds without positive seasons. They are also worlds without much warmth, light or oxygen, with a lot of moisture, without much to eat and without most of the traditional patterns of life. It may be that lack of oxygen, resulting both in sluggish metabolism and slow development, is more responsible for all-the-year round breeding than the relative lack of seasons; perhaps not. These cave worlds, devoid of nights and days, of summers and winters, ought to be revealing. But they are not. At least, not yet.

13 The Sexual Seasons

The first twelve chapters of this book, save for the forward-looking Chapter 3, have described various aspects of the world's seasons and the world's climate, and have concluded with a brief questioning look at a world almost without seasons. The activities of animals and plants have been mentioned, but have not been discussed. They have been observed rather than inspected. The second half of this book examines their various adaptations far more closely. As the paramount reaction of life to a seasonal existence is the manner of its sexual cycle, this forms the first subject of the book's second half.

That there are nesting seasons and breeding seasons and sexual seasons is undeniable. It is equally axiomatic that there are advantages to these seasons. What nestling would survive if its parents could not find sufficient food at that time? It is also true that breeding seasons are, for almost all species, in keeping with the seasonal cycles. For example, with clockwork regularity very different from the erratic timekeeping of normal climatic conditions, the English starling lays its first eggs in the last two weeks of April. These are hatched in May, a month suitably abundant with food, but the process had been initiated long beforehand. Just as a Christmas card manufacturer does not suddenly leap into activity during December, so does all biological timing work by anticipation of the future. The sexual season of mating occurs before the eggs are laid, and they are laid before being hatched, and they are hatched before the offspring learn to fly, and the offspring must fly before they can migrate, and of course they must migrate before the cold weather arrives, before food vanishes and the environment turns lethal. These various cycles, all a part of the major cycle of reproductive life, are usually in tune with the astronomical cycle of the Earth going round the sun.

There are exceptions. In the spring a young man's fancy turns to thoughts of love but, as Bernard Shaw added, it also happens in the autumn. He was right because man and the apes form a particular group which breeds all the year round. Similarly, man and various other occupants of his heated homes, such as rats and mice, can take advantage of the eternal presence of food and warmth, and their winter offspring will fare almost as well as summer progeny. (But not quite. There are subtle differences affecting even man's well-regulated existence. Neonatal mortalities vary with the seasons. So does birth rate. So does eventual mortality, for the disease of death can sometimes be linked to the season of birth occurring all those years beforehand.)

The dominant complication of the sexual cycle, irrelevant to some of the smaller animals, is that sexual activity cannot take place in the spring if the progeny are to be produced in the spring. For one thing, the gestation length has to be allowed for between mating and birth. In some creatures, such as the horse and the wildebeest, this is happily just under twelve months. Mating and delivery can therefore both be at the same time of year, and during the same season. (This sort of timing makes one wonder what differences to expect in the animal kingdom if the year were of quite a different duration, perhaps more like the 225-day year of Venus, or even like the 687-day year of Mars.)

In many animals, such as most small rodents, there is less seasonal complication because the gestation time is so short. With, say, only three weeks between copulation and birth, and then only, say, six weeks (for the mouse) before the young are themselves pregnant, a single warm season can witness considerable reproductive activity and many families. Using the spring as a mating season, whether for near-immediate delivery or for delivery one year later, is satisfactory provided the gestation length is long enough or short enough to permit such a system. In certain intermediate instances the gestation length itself has been substantially modified, in Procrustean fashion, to make it fit. Admittedly all gestation lengths are subject to the customary pressures of natural selection, but the phenomenon of delayed implantation is something quite different.

If pregnancy is overlong, and cannot become shorter, it might as well be made yet longer. This, loosely, is the principle of delayed implantation, and seals, some deer, many marsupials and various members of the stoat and weasel family *(Mustelidae)* are all examples of this phenomenon. The gap in time between their mating and the delivery of their progeny is too long to be justified by the foetal development itself. Rather than slow down the whole process of development, which is plainly an extremely fundamental undertaking, the system of delayed implantation involves a stop-start procedure. Fertilization takes place, development begins, and then it stops. Later it begins again, eventually to proceed at a normal pace. (There is argument about the term delayed implantation. Suspended development has been proposed but is still far from supplanting its predecessor. Currently in favour among workers in this field is embryonic diapause.)

With the grey seal the total gestation time from copulation to delivery is 351 days. Following fertilization, normal development proceeds until a small blastocyst is formed, a stage reached in comparable animals after about six days. Then, unlike the comparable animals, virtually nothing happens to the seal blastocyst for the next 100 days or more. Perhaps there is a total of four days' worth of normal development throughout these three months. Proper development then starts again, slowly at first, and finally at the conventional pace. In other words, fertilization occurs about two weeks after the birth of the previous pup and the next young one is born eleven and a half months later rather than eight months later. Eight months would be an awkward interval from the point of view of the seal habit of returning to breed for only a brief portion of the year. The advantages of a delaying system are that the mating can take place during the gathering of the seals for the annual birth of their pups, and also that the mother is not having to suckle one active young pup while major demands are being made internally via the placenta for the next pup. At least next year's offering is quiet and undemanding for those 100 days. It waits its turn.

Admittedly dogs are litter animals while seals produce singletons, and litter amimals suffer competition within the uterus enforcing an urgency hurrying up birth, but it is interesting that birth can be such a moveable feast (to use Sir Peter Medawar's phrase). A large dog produces its pups in nine weeks. A small seal, like that grey seal, so delays things that its pup is born after fifty weeks. One might have assumed that dogs and seals, somewhat similar animals in many respects, would have similar gestation times, but it is not so. Even the leisurely human being, where all the emphasis is upon slow maturation, only takes thirty-eight weeks between fertilization and birth, or twelve weeks less than the seal.

This phenomenon of arrested gestation was first described in roe deer. They mate in late summer and give birth in May although the real gestation period, when development is actually happening, is nearer five months than the apparent nine. Perhaps the phenomenon seems even more striking in the smaller animals because they customarily take less time to produce their offspring. One might have suspected that bats would have a short gestation, but many copulate in the autumn and do not give birth until the high summer of the following year. For them delayed implantation is much less apt a term because even fertilization does not take place for many months following copulation. The donated sperm are kept throughout the winter, perhaps in the uterus, and they only fertilize the eggs when these have been produced in April. Development is then swift, and the young are born in June or July.

It scarcely needs underlining that spring is a good time in which to be born, but it does need to be emphasized that daylength can be far more crucial than warmth for a successful rearing of the brood. Warmth is necessary in temperate climates to create the upsurge of vegetation and of insects, but daylength is necessary to take advantage of such bounty. In this regard the temperate summer has much more to offer than the tropical world where maximum daylength is, by definition, never more than thirteen and a half hours. At the polar world it can be twenty-four hours, and the temperate world is the happy compromise with both plenty of food and plenty of daylight to find it in.

This benefit of latitude has been demonstrated by observation of two pairs of north American robins, one pair living in Ohio, the other in Alaska. On average the northern birds could fit in 137 feeding visits per day whereas the Ohio birds could manage only ninety-six visits. This meant that the northerly nestlings were fit enough to leave the nest in nine days as against thirteen days in Ohio. Therefore, not only do the young develop faster, but the vulnerable period of a defenceless nestling, without the means of escape from danger, is much reduced. To emphasize this point even more strongly, although natural history is plagued with exceptions, birds do not nest and bring up their young whenever or wherever the daylight lasts for less than eleven hours.

To live far from the equator is a compromise. The days are longer and competition is less, but the winter is more severe and food is less abundant. One result of the conflicting benefits and disadvantages is that the egg-laying season for each particular bird is steadily delayed with increasing latitude. Moorhens and coots (the family *Fulicariae)* have a wide range, and prove this point. The farther north they live the later the breeding, with the only complication being that there may be time enough for two broods within the tropics. Apart from this exception the coot egg-laying season starts twenty to thirty days later for every 10 degrees of latitude from the equator. Such a reduction in available summertime cannot, of course, go on for ever, and the *Fulicariae* do not breed north of 60° N.

Such precision over timing, and such an association between latitude (or daylength) and the breeding season, correctly indicate an effective system for appreciating the seasons of the year. For some reason most humans totally forget about shifting daylength until the season is so advanced, say a couple of months after the winter solstice, that they are forced to admit the days are drawing out. Animals can be far more sensitive. Deer, for example, are generally in rut within a week of the passing of the summer solstice, although this may be a delayed response to the long but still lengthening days just before midsummer rather than an immediate response to the post-midsummer shortening. Even so, what human being without a clock could hope to know within a few days when the nights first start to draw in? And what human being without a calendar could hope to know when the year had reached, say, the first week in February?

A good illustration of human inferiority in this matter is the fact that certain human societies do use animals as clocks, and as a calendar. No example is more renowned than the palolo worm *Eunice viridis,* a polychaete which lives in rocky crevices on the shorelines of Samoa and Fiji. For most of the year it lies hidden,

Cave life Neptune's Cave, Sardinia. Caves are without night and day and have little seasonal variation: annual temperature and humidity variation may be as little as 30°F (−1°C) and 4%. All-year-round breeding among cave creatures may, however, be due to lack of oxygen and the resulting slowing down of the metabolism as much as to the lack of seasonal stimulus.

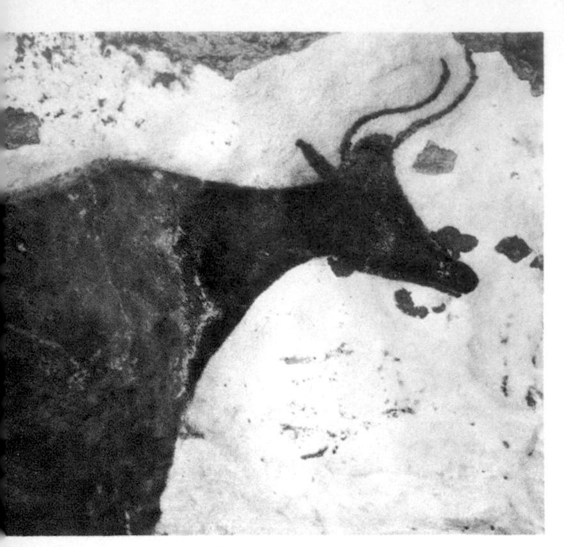

RIGHT: Enlarged examples of cave insect life. Such forms of life normally lack pigmentation and vision, have atrophied respiratory systems and take in oxygen from the moist air through their bodies. Cave life can, however, be congenial to man, providing a refuge from temperature extremes. During the ice age, both men and animals went underground.
ABOVE: One of the famous palaeolithic cave paintings at Lascaux, France.
BELOW: Pueblo Indians' cave dwellings at Mesa Verde.

ABOVE: The South American guacharo bird breeds mainly in caves.
LEFT: A roost of dog-faced bats in the Batu caves, Malaya. Bats live in caves throughout the world—the largest known cave, the Carlsbad Cavern, was discovered after a cloud of millions of bats had been seen emerging from the ground.
BELOW: Blind white cave fish. Blindness and whiteness are typical attributes of cave water forms.

Sexual cycles Gestation is an important factor in determining sexual cycles: the best time to mate is determined in accordance with the best time to be born, except in the case of men and apes for whom, on the whole, any time is a good time. Growing daylength may act as a sexual stimulant (in ferrets, hedgehogs, racoons); so may decreasing daylight (sheep and goats). Amphibia react to temperature change; the motivation of invertebrates is more complex and in many cases remains a mystery (the Pacific palolo worm reacts to the phases of the moon). The pituitary gland, which is the principal hormone producer, is the main factor in the reproduction cycle of animals: much work remains to be done in this sphere.

ABOVE: A case of spring madness in March hares. Spring, however, is not the only mating time: the stag (TOP CENTRE) comes into rut in late summer.

The physical appearance of animals is often adapted to sexual ends, sometimes with extravagant results as in the case of the mandrill monkey (TOP LEFT) and the lyre bird (CENTRE).

Sexual needs often determine group behaviour: BOTTOM LEFT: A colony of mating gannets. Some animals come together in groups to mate; others prefer seclusion.

Birth Birth can occur in any season but animal gestation is sometimes arrested so that the young can be born at a favourable time of the year. The pup of the grey seal (BOTTOM RIGHT) is born after a total of fifty weeks to avoid the winter period.
ABOVE: A vixen with her fourteen-day-old cubs, born in spring.

TOP RIGHT: A baby kangaroo in its mother's pouch. Like all marsupials the kangaroo is tiny at birth, measuring under an inch. It stays in the pouch until old enough to run.

OVERLEAF: Young egret straining for regurgitated food from the mother bird.

seemingly unaware that anything is happening beyond its particular fissure. Then, unfailingly in October and November, it spawns. The precise moment of spawning is at dawn, both on the day before and on the very day when the moon enters

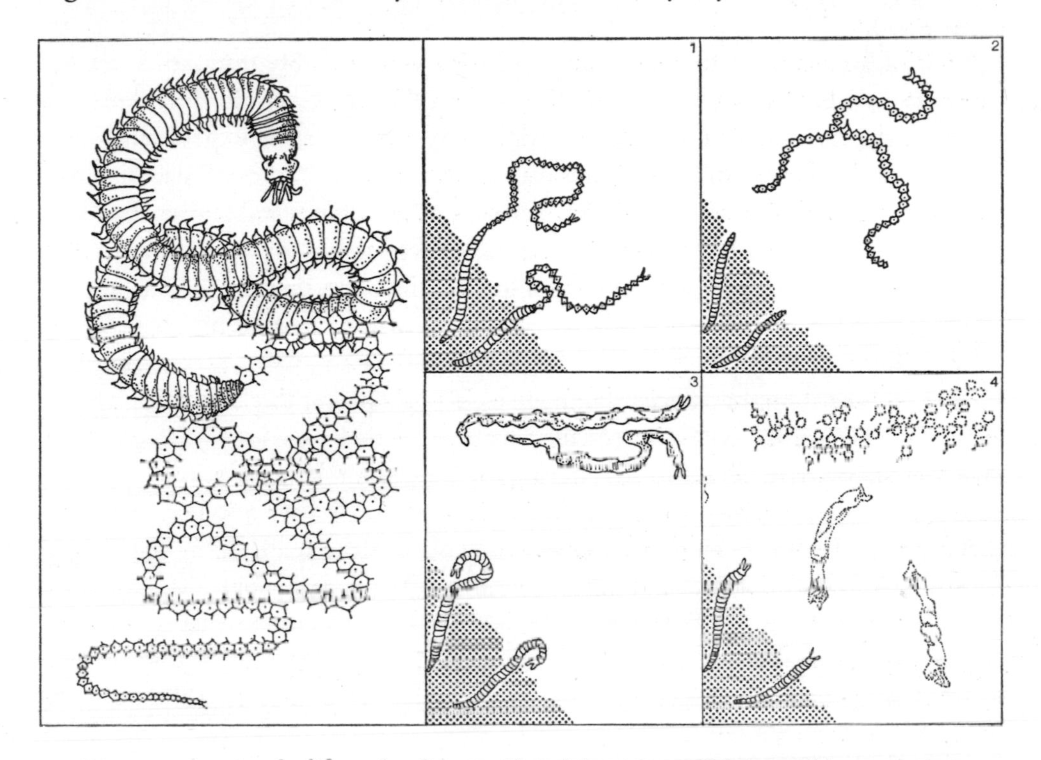

15. Diagrams showing the life-cycle of the Pacific Palolo worm, which invariably spawns in October and November on the dawn of the day before, and of the very day when the moon enters its last quarter. (1) The worms reverse their position in burrows, the back half altering drastically as the reproductive organs grow. (2) This part breaks free and swims to the surface. (3) The remaining adult portions reverse to a normal position. The eggs and sperm section develop further with internal segments breaking down to give a single sac. (4) This then bursts releasing eggs and sperms, resulting in random fertilisation – is so timed that millions of worms are all mating at once.

its last quarter. Having arisen in such a precise manner to reproduce its kind this Pacific worm then subsides until the right time of the right phase of the right moon on the following year.

The worm's incredible skill is related not only to the season but also to the moon, thereby successfully combining both astronomical events in a way which mankind has tried to do with months and years. Most animals are principally dominated by the time of year – warmth, daylength, vegetation – rather than the phases of the moon. Mankind, hypnotically intrigued by the moon's timing and human ovulation, even named menstruation after the moon in many of his languages, but despite all efforts to discover parallels between moon phases and the menstrual cycle, none have yet been discovered. Indeed if the moon is linked with menstruation, as James Joyce wrote with happy simplicity, 'why don't all women menstruate at once?'

Lack of human interest in changing daylength was probably responsible for the late date on which it was first scientifically shown that light was of primary importance in the rhythm of the breeding cycle. The all-important experiment (on a junco, described more fully later) was carried out in 1924. Although scientists had failed to absorb this point beforehand, others in other spheres had known that daylength was vital. The Japanese, for example, frequently made pet birds sing throughout the winter by exposing them to artificial light, thereby influencing the sexual cycle. Hens were made to lay deep into the winter months by giving them more light.

Had anyone thought sufficiently about it, it is plain that daylength has a constancy which is not necessarily equalled by any of the vagaries of climate. The swallow may arrive in mid-April during an unseasonable snowstorm and after weeks of similar weather, or it may arrive in mid-April after a long and balmy early spring. The bird's clockwork consistency is therefore matched by the Earth's progression round the sun, and not by the myriad and perplexing cyclonic progressions around the world. Hours of daylight are reliable and weather is not, save in the most general terms. As temperature is crucial to vegetation, it was perhaps this fact more than any other which caused the assumption that temperature must be the keynote for the entire animal world.

The experiments in 1924 stimulated other scientists into further discoveries. In 1932 the first study concerning light was made on a mammal. Reproduction virtually ceased in the field vole/mouse *(Microtus agrestis)* when its daylength was shortened from fifteen to nine hours. Since then it has been learned that an animal's sexual cycle is set into operation after its annual quiescent period by a wide variety of factors, quite apart from mere length of day. For example, gradually *increasing* daylength acts as the trigger for many species, such as the ferret, the hedgehog, the raccoon and the regular starlings with their first eggs in the last half of April, but the trigger is *decreasing* daylength for sheep and goats. Early ferret experiments were made more complex, and more fascinating, by the discovery that the male remained infertile if the increase in daylength was given all at once. The trigger of more light was only effective if the additions came in small increments at a time.

Whereas light is so important to mammals and birds, temperature is understandably more important to those animals relying upon outside warmth for their activity. A few reptiles have been found sensitive to light as a trigger for sexual development, but no amphibia. Temperature is generally the crucial factor for the amphibia, but rainfall can also be important. (Plainly the amphibia requirement for water, even more important for the eggs and young, demands that reproductive activity must be in association with wetness, and therefore rain, however right the time of year may be.) With fishes the fact that it is raining is obviously of little consequence, and temperature is generally the vital primary stimulus. The right temperature may be inadequate by itself, and often there has to be the right light as well.

The invertebrates have been far less effectively studied. One would suspect that temperature in the cold-blooded invertebrate world is of the greatest importance, and certainly more so than light. But what about that palolo worm? What can a mechanism be that responds to the situation just before dawn in the last quarter of the October and November moons? And what is so different about these two moons as seen from the equatorial, ever-warm waters of the South Pacific? The palolo worm is never wrong.

John R. Baker of Oxford University, cut a clean swathe through much of the complexity of this subject when, with very good sense, he divided the causes of a breeding season into two. He wished to distinguish between the ultimate cause of a particular cycle running at a particular time, and the immediate reason why that cycle actually takes place at that time. There is a very real distinction here. The ultimate cause is long-term in that there are, say, plenty of insects for food during a particular season. The more immediate or proximate cause (Baker's term) is the increasing daylength of early spring which sets the cycle in motion, but the long days of summer may be the ultimate cause of breeding, in that the number of feeding hours in every twenty-four can then be considerable. It is important to bear Baker's distinction in mind when thinking of causes for any particular cycle.

Ultimate causes in the tropics are hard to deduce, especially when every season seems ripe for breeding, but it should be remembered that overbreeding can be a problem and it is not axiomatic that an animal *must* always breed when it can. Natural selection by no means favours species merely because of their abundance; large numbers can lead to food lack, and this can lead to extinction. It has therefore been argued that a need exists in the tropics for a restricted breeding season just as much as in other parts of the world.

The proximate causes of breeding in the tropics are also rather more subtle than elsewhere. The important trigger of light is less seasonal there. Light intensity rather than duration may act as a replacement because most tropical seasons vary in their cloudiness, and therefore in their light intensity, as they build up and then pour forth their bouts of rain. Or perhaps the trigger is not so much light in general as its ultra-violet border in particular which also oscillates in intensity according to the time of year. However subtle the causes may be, reproduction in the tropics is invariably seasonal.

Exceptions to this rule are very few. The fantail warbler of Madagascar, for example, breeds throughout the year, but even it has a breeding season in other parts of its range. Strict adherents to the rule are not only more numerous, but are frequently astounding. There is the tropical bat *(Miniopterus australis)* which spends all day in a dark cave, and whose females become pregnant only once a year at the beginning of September. The regularity with which these and other animals adapt their ways to include a definite breeding season, even when the seasons are so indefinite, assists in the conclusion that animals' breeding seasons are extremely beneficial to animals, however slight the climatic seasons that trigger them.

The proximate and ultimate causes for breeding at a particular period, or periods, of the year are certainly important. Nevertheless they coexist with the animal's basic internal rhythm, whose length of cycle is usually about one year. Without this fundamental attribute the proximate triggers would have nothing to fire, no internal charge to work on. It is this basic rhythm which, given suitable stimulation, causes the order of things – the preparation of the gonads, the courtship displays, the mating, the hatching, the sexual inactivity, and then the starting up again of that activity. Without the basic rhythm being in a suitably receptive phase the triggers are powerless. For example, swallows flying to Europe in the northern spring encounter lengthening days, abundant food, empty niches and a situation suitable for breeding. So they breed. On the other hand, swallows returning from Europe in the northern autumn cross the equator, and again encounter lengthening days, plenty of food, good niches and circumstances favourable for breeding, and yet they do not breed. The internal cycle is only prepared to have one breeding season a year, not two, however appealing the circumstances may seem to be. There is nothing suitable for the trigger to fire, and so nothing happens.

As a parenthesis to the main story it should be pointed out that the aptly named seasonal isolation must often have been a potent force in the development of new species. Supposing that, for some reason, half of a particular species go through their sexual cycle a month before the remaining 50% of individuals go through the similar procedure. If the reasons for earliness and lateness are genetic and inherited this process will continue. The early birds and the late birds may meet, but they will never mate. Therefore any differences arising between them, any mutations, will not be shared by both. In time the differences may be so great that two species will have resulted, the early breeders and the late breeders. Then, even if a contrary year does trigger them both into breeding together, attempts at cross-fertilization would be useless, and the two species would remain distinct. Their seasonal isolation would have had its say. It alone would have caused the existence of the two distinct species.

Underlying every statement in this chapter, however unobtrusively, has been the endocrine system, the glandular control mechanism which both causes the cycle and which is itself caused to function by the cycle. Therefore the endocrines will have to be examined. However, to prepare the ground still further, and to describe the cross-linkages involved even in the elementals of a breeding season, the subject ought first to be looked at from the evolutionary point of view. The fact that natural selection ordains this or that feature is fundamental. It does influence, it does control, and it does its job on a multiplicity of factors in a style that only a computer's brain can fully absorb. Mere human brains need to dwell upon this point for a moment, and determine what is meant when an animal is said to breed at a certain time because it is advantageous to do so.

To migrate north, for example, into the northern hemisphere involves a long trip, hazards en route and the consumption of energy. It eventually means arrival in a world of longer days (certainly), lower temperatures (probably), less abundance of food (probably), but less competition for that food (probably). There are also likely to be fewer resident predators. On the other hand there is less time, in terms of weeks, less opportunity for more than one brood, and always the urgent need to migrate south before the arrival of short days, harsh weather and cold nights. The one brood produced, or two or more, is also subject to natural selection. The respective advantages of large broods, coupled with intra-uterine and intra-nestling competition, always produced in a hurry and probably with large losses, have to be set against the advantage of small, well cared for, well nurtured and high survival offspring. Is the gestation long or short? If long, will it take long to build up the population after each setback? The average birth rate is yet another compromise between respective merits, between the arrival of warm days, of enough food, of enough time. When will the young be mature? Will they change their colouring/fur/plumage as they develop? Will they mature in time to produce more offspring of their own that season? Is such haste, with inevitable losses on the way, an advantage?

In short, there is a wealth of compromise, of the comparison of unequal assets. The final result is certainly not an ideal situation. People go around saying that nature knows best, that this particular bird has found the optimum solution to its particular requirements. It has done nothing of the sort. Instead its life is a perpetual compromise, a temporary state of affairs which has enabled it to survive, but which could certainly be improved. Natural selection always has to make do with the material at its command, plus the occasional mutation, and the material is a collection of left-overs, of atavisms, of remnants from an earlier age when these too were compromises from a yet earlier assortment of characteristics.

Mankind, boggling at all the interactions in nature and being incapable of distinguishing between so many rival benefits, happily relaxes in the view that natural selection is all-powerful. We say it must be advantageous to migrate north, despite the disadvantages, for otherwise the birds would not fly north. Such action would have been selected against had it been unsatisfactory on balance. The process of evolution surely weighs up the counteracting influences, and the resultant behaviour and practice of each species is the compromise result. Natural selection can only modify rather than alter, amend rather than substitute, and it produces evolution rather than revolution. Nowhere is this more obvious than with the muddled up assortment of endocrines, each stimulating and being stimulated in turn, each a mixture of singularity and interdependence. The bewilderment of breeding seasons cannot be thought of without remembering the hormonal bewilderment at the back of it.

Hormones, also known as messengers, are from the Greek word to excite, and the modern term was introduced early this century. Later it was realized that hormones both inhibit and excite, but the word remained. Its original introducer (E. H. Starling of University College, London) defined a hormone as any substance normally produced in the cells of some part of the body and carried by the blood stream to distant parts which it affects for the good of the body as a whole. But this definition suggests that animals without a vascular system cannot have hormones, and Julian Huxley subsequently produced a more liberal definition: a hormone is a chemical substance produced by one tissue with the primary function of exerting a specific effect of functional value on another tissue.

Among the vertebrates the pituitary gland is the principal hormone-producer and master key of the reproductive cycle. It looks very different in many of the vertebrate species, but in general it is egg-shaped. It is always situated in the centre of the base of the skull, attached by a short stalk to the base of the brain. Although small, and seemingly homogenous (at least in animals higher than the lamprey group), it is always of two distinct parts (the lampreys have them actually apart from each other) and these are called anterior and posterior. It is the anterior lobe which produces the hormones of most immediate importance to the reproductive cycle, some of which act on other endocrine glands and some of which act more directly.

A direct effect is upon the egg follicles and the sperm-producing cells. It is this gonad maturer which is known as FSH or follicle-stimulating hormone (and it is this hormone which has been achieving renown recently when its injection into certain previously infertile women has caused them to have either single births, or, infrequently, as many as six at a time.) Another gonad stimulant is LH, or luteinizing hormone, which affects ovulation and the corpus luteum, the immediate successor to the egg follicle. LH has other less positive effects upon both the ovaries and the testes, and thereby complicates the issue by acquiring another name, ICSH, or interstitial-cell-stimulating hormone. Still a third hormone from the pituitary, known by yet more initials, namely LTH for luteotrophin, acts upon the corpus luteum to make it secrete the hormone of pregnancy, namely progesterone. It is not just the plethora of initials and names which is confusing; it is the number of effects and counter-effects.

As an example of this bewildering endocrine interaction, here is an abridged quotation from the work *Animal Hormones* by J. Lee and F. G. W. Knowles, where they describe the probable cycle of events of a generalized sexual cycle. The anterior pituitary

> secretes FSH in increasing amounts and this hormone induces ripening of the ovarian follicle which in turn secretes oestradiol. Oestradiol acts on the anterior pituitary directly, or indirectly, to decrease the amount of FSH secreted. If the level of FSH falls it is suggested that LH will be secreted . . . After ovulation blood levels of LH increase which induce the development of a corpus luteum, but it may be that this will only secrete progesterone if sufficient LTH reaches it. As the corpus luteum matures, increasing amounts of progesterone are produced which depress the output of LH and LTH . . . If the output of LH and LTH falls, the anterior pituitary will secrete FSH and another follicle is stimulated.

In other words, the elementary cycle of ovulation, which produces one egg after another, involves the very minimum of five hormones, three of which stimulate development and two of which act by feedback to depress a particular hormone production.

To add once again to the confusion there are hormones produced by the placenta which are similar in their effects to the pituitary hormones. Of course the placental hormones can only be produced during a pregnancy, when a placenta actually exists, and the hormones are just called APLH (or anterior-pituitary-like-hormones) for convenience. The placenta, it seems, is designed to look after itself as soon as it can. The pituitary's hormones are essential to the maintenance of pregnancy, but pregnancy will continue even if the pituitary is cut out for APLH will take over.

The testis, the only sperm producer, is also a hormone producer once it has itself been stimulated into activity by other hormones. All substances responsible for the development and maintenance of maleness are called androgens, and the particular androgen produced in the mammalian testis is testosterone. (Oddly the testis also produces oestrogen, the female hormone, thus illustrating the point that the difference between the sexes is principally one of degree, not of kind. Similarly, the mammalian ovary produces androgens as well as oestrogens.) Testosterone stimulates the development of male organs as well as the secondary sexual characters of maleness, such as facial hair, the longer larynx, the stronger muscles, and various masculine aspects of behaviour. Of course, secondary sexual characters vary enormously between species, what with antlers, horns, colour, plumage, and manes as well as moustaches, but testosterone is the principal hormone behind them all.

Most vertebrate males remain sexually potent whether their females are having receptive cycles or not, but various other males have annual cycles of their own, only coming into rut, for example, once a year. The production of testosterone is therefore constant or varied according to the animal. Also its effects vary according to the current hormone situation; it is not acting in isolation, and the fluctuating production of these other hormones can cause the effects of testosterone to fluctuate in parallel, even though the actual production of testosterone may be more or less constant throughout the year. (The interdependence of hormones is the critical feature of the sexual cycle, and of the endocrine system in general. Nothing hormonal should therefore be examined by itself, but a start has to be made somewhere and it is easiest to consider the hormones one by one as if independence existed. Testosterone is just one such hormone.)

The ovary, despite its totally different and more complex role, is similar to the testis in that it also secretes hormones as previously noted. The female equivalents of the male androgens are the oestrogens (generally oestradiol) responsible for the development of the accessory female organs and the female's secondary sexual characters. The accessory female organs are those which are positively linked with the sexual system. The secondary sexual characters are not considered part of the sexual system, however important the part they may play in the sexual cycle.

Many vertebrate and all mammalian ovaries also produce the hormone progesterone. This co-operates with oestrogen in producing its effects, but it also helps in the establishment of a pregnancy after a successful fertilization. The male hormones produced by the ovary are quite as inexplicable as those female hormones produced by the testis. Both remain enigmatic, and particularly so when abnormal quantities are being produced, so that a hen, for example, develops a cock's comb and wattles.

There are many other hormone producers in the body, such as the adrenals, the thyroid, the parathyroid, the pancreas, and the various endocrines of the alimentary canal, but the hormones already mentioned are those most intimately concerned with the sexual system. They are the principal effectors of the cycle which is triggered from the outside, organized by the hormones from the inside, and which is then observed from the outside as courtship, mating, breeding, and replenishment of the Earth.

Two further general hormonal points concern their mode of behaviour. They work principally by stimulating the development of specific organs, and their life is generally short. Quite how they activate this organ as against that similar nearby organ is unknown, but they are certainly able to do so. They also disappear rapidly, either into the urine or by transformation into some other chemical. This point can be proved by the injection of a hormone which is not being produced by the animal at that time. For a short while it will be detectable in the bloodstream, but thereafter it will have vanished. Hormones have to be in steady supply if their effects are to be kept up. Without such a constant production their possible effect will, probably, be countermanded by some other hormone.

It is now important to describe, in rather more detail than earlier, the round of hormonal events in an ordinary oestrous cycle. That earlier quotation (from Lee and Knowles) demonstrated the inherent interdependence of the system, and this point is even more emphatically made if the sexual cycle is delved into yet deeper. The easiest place to start is the change over from the inactive sexual time to the active, a moment customarily associated with springtime, but which has to have its beginnings distinctly earlier.

The cycle is set rolling in the female by the sudden activity of the anterior pituitary. Something has to trigger off this initiation, such as the arrival of the appropriate season, or merely the adulthood of the individual; both such triggers will be referred to later on. Anyhow, the pituitary pours out its FSH, the ovary then aids and abets by its production of oestrogen, and both cause the ovary's follicles to grow and then to burst. The ovary's oestrogen is soon secreted to such an extent that it is able to depress the pituitary and reduce its production of FSH. At the same sort of time the production of ICSH by the pituitary is stepped up, a change probably caused by the quantities of oestrogen then being produced. The ovarian follicle, as explained beforehand, becomes the corpus luteum, and this successor to the follicle secretes the hormone progesterone. This new hormone then further stimulates the preparation going on in the uterus. It also stops any further production of FSH by the pituitary, and it encourages, somehow, the body's removal of any remaining oestrogen by the kidneys.

Assuming no pregnancy, the corpus luteum then becomes less and less active and its supply of progesterone drops off. Following this reduction there exists, to speak politically, a power vacuum. The progesterone is declining, the oestrogen has been removed, and the balance of power is no longer controlled by these two. Consequently, with both out of the way, the anterior pituitary can flourish again and produce its FSH once more. The political analogy still holds because the pituitary steps into the breach so forcefully that it starts up the whole cycle once again. In this case, unlike the first oestrous cycle of the season, the trigger is merely the abrupt absence of restraining forces.

The annual cycle is basically similar to each oestrous cycle running within it, save that the period of anoestrous, the negative time, is the dominant feature of the non-breeding season of the year. During anoestrous the anterior lobe of the pituitary shrinks in size, none of its sexual hormones can be detected, and everything is quiescent – until spring comes. Then having been prompted by whatsoever feature of the changing year does the trick, the anterior lobe grows anew, it produces many of its so-called basophil cells, and soon the existence of its hormones proves that the cycle is once more under way. The testes and ovaries are stimulated by these hormones, and their own hormones, testosterone and oestradial, are soon in production. The sex organs then grow, the secondary characteristics manifest themselves more plainly, and breeding takes place.

Whether this happens, once, twice or many times a year, or even every other year, depends upon the species, although the fundamental pattern remains the same. In its essence the annual sexual cycle is one magnified oestrous cycle. Menstruation, the human and ape system, is basically no different because the hormonal cycle follows the same pattern. The menstrual cycle has been given a different name from the oestrous cycle because oestrus (in heat, etc.) and ovulation occur at the same time whereas menstruation (either the start or the end of it) and ovulation do not. Although this is an important distinction, the basic hormonal control relating to both systems is identical.

Whereas the effect of a hormone upon an organ is direct, in that the hormone is wafted through the organ with the blood, and therefore inevitably encounters the cells to be stimulated, the effect of the season upon the hormone glands might seem more obscure. There is, for example, the intermediate stage of the sensory system, for each external seasonal trigger of an event has to be transmitted via one or more sense organs to the appropriate gland or glands. The testing of the effect of a hormone upon an organ is fairly easy, because the artificial injection of that particular hormone soon supplies an answer. The experimental procedure for discovering the link between, say, light falling on the retina and the subsequent activation of a gland has been far less simple to establish.

As so many animals make use of light, whether of increasing or decreasing daylength, or even of quantity, it provides a valid example of the problems involved. So far, not all the links in this problematical chain of events have been unravelled. No one knows the entire mechanism whereby increasing sunlight leads, for example, to increased FSH. However, it is at least known that there is no direct simplicity, for there is no direct nervous pathway leading from the retina and along the optic nerve to the pituitary gland.

In fact the system is far less simple, for it appears that there is not even an indirect nervous pathway leading from the retina to the gland. The principal stumbling-blocks to any such notion are not only that there are no nervous connections between the visual areas of the brain and the pituitary but that there are no nervous links between the hypothalamus and the anterior lobe of the pituitary, namely between that part of the brain through which the links would have to pass to reach the pituitary and that part of the gland which is producing the right sexual hormones. At best the nervous control of the anterior lobe by the hypothalamus would have to be of some indirect kind, such as constricting or enlarging the blood vessels supplying the gland with its essential nutrients.

Consequently, as nerves are out, in the sense that the correct nervous pathways do not appear to exist, the stimulation of the gland via the retina has to be non-nervous somewhere along the line. If some substance were secreted at one point, and it was discovered that this secretion duly affected the gland, the gap in the pathway would have been explained. The hypothalamus is very much the master-mind of many basic activities of the body, as it is involved in sleeping, waking, appetite, thirst, metabolic and temperature regulation, to name a few – as well as with the autonomic nervous system itself, the self-governing network of control. Therefore it is easy to postulate that the hypothalamus is also a secretor, and must be involved somewhere along the complicated line of the indirect stimulation, by light, of a sexual organ. In fact evidence has arisen to justify this presumption, for not only does the anterior pituitary start to secrete hormones when the hypothalamus is experimentally stimulated, but there are minute venous channels connecting the anterior lobe with part of the stalk connecting the gland to the brain. So far so good, and extremely encouraging. Unfortunately, although the hypothalamus is now known to produce a secretion in addition to all its other duties, the principal hypothalamic secretion known to affect the pituitary gland affects its posterior lobe. Therefore either the posterior lobe affects the anterior lobe in some, as yet undescribed, manner or the hypothalamus secretes some other, as yet undetected, hormone. Somehow the chain of events has to link the eye and the pituitary for organisms whose reproductive cycles respond mainly to seasonal changes in light.

The fact that the hypothalamus *and* both the pituitary's lobes are all so intimately crowded together at the base of the brain does not mean that the stimulation of one inevitably leads to the stimulation of the others. There has to be a connection of some kind, whether nervous or hormonal, and that connection has yet to be unravelled. In fact, the discovery of minute pathways and new hormonal substances, passing perhaps between two halves of an even smaller gland, is the crucial and perplexing nub of that grandiose earlier statement which simply asserted that changing light causes changing hormonal output. The finer points of that remark are still most incompletely drawn.

In short. taking all the points together, the evolution of breeding seasons, their various triggers, their adaptations (such as delayed pregnancy), their mechanisms, and their systems of control, the annual sexual cycle is a highly complex affair, the ramifications of which are – to say the least – only partly understood. Mankind must have known about the sexual seasons of life ever since the word mankind was applicable, ever since a man could consciously take note of anything, but only very recently has any real investigation of the subject been made.

Only in 1924 was it first proved that changing daylength acted as a trigger. Only in the 1930s did John Baker do his pioneer work and original thinking on the subject. Only very recently has the role of the hypothalamus been suspected and suggested, and most of the unravelling has yet to be done. Much of the excitement of science lies not in its discoveries, but in its initial hints that there is far more to a particular feature than had at first been supposed. Certainly there is far more to a sexual season than had ever been dreamed of by most of our forefathers, however much they noted when the whitethroat sang or all the swallows arrived.

Therefore, it is reasonable to remember the apparent simplicity of one sexual season, but to have at the back of one's mind the magnitude of physiological involvement which such simplicity entails. For example, let the sexual season be that of a bird. Let the bird be a temperate dweller which neither migrates nor hibernates, and let there be nothing outstanding about its annual cycle to merit amazement. In January, with gonads still fairly undeveloped, it starts defending a particular territory. In February it pairs up with one of its own species. In March, with gonads ripe, it mates, and fertilization is effective. In April it makes a nest and lays eggs. In May incubation leads to hatching. In June the nestlings are fed. In July they fly, and both they and the adults are equally infertile at that time, with their gonads in a similarly undeveloped state. In August all the birds part company, and they lead independent lives in other areas. In late autumn the gonads begin a recrudescence. By the end of the year they are growing rapidly, although the days are darkest and the climate, if not at its most unpleasant, soon will be.

It all appears at once so simple, so clear cut, and yet so bafflingly complex, so interdependent. In addition there are those two great adaptations to the seasonal year, namely migration and hibernation, and they do nothing to simplify the story. In fact each deserves a chapter to itself.

14 Migration

Migration is, at first sight, a totally explicable response to the changing seasons. The swallow arrives in England during April, when daylength is growing rapidly and the days themselves are becoming warm. In autumn that same (British) swallow then hurtles back with its brood to the eastern part of southern Africa. Quite obviously, or so we say to ourselves, the bird takes advantage of the temperate summer in the northern hemisphere and then of the sub-tropical summer of Mozambique, Natal and the Cape. It is undoubtedly getting the best of both hemispheres.

Then doubt arises. Why all that way? The loss of life en route must be appreciable, the consumption of energy prodigious. Even tourists know it is not necessary to travel 7,000 miles to escape the English winter. In fact the swallows from Germany tend to stop relatively short, in the Cameroons and on the Congo river, but even they are making a formidable journey twice a year. Still other swallows, emanating from Europe, regularly winter in North Africa. Surely, as so many of the wealthier dowagers would agree, that is far enough? Even more incomprehensible are the migrations from which hardly an animal returns, as with the lemmings.

The urge to migrate is undeniable, and succeeding generations show the same trait; yet how is it inherited if the emigrants themselves never return? And what is the advantage? When the losses are virtually 100%, as with the lemmings, does this fact help to explain the lesser losses of other migrations, of the thousands of geese coming to grief high up on the Himalayas, of the bison careering onwards to be drowned in the rivers or killed by the precipices? (In 1858 over a hundred thousand carcasses were found at the foot of one cliff, the folly wholly self-induced for not a single wolf carcass was found with them.) Losses are undeniable; it is the gains that are so supect.

When a species loses thousands of its number, or migrates from some congenial area into one of hardship, or from one region to a similar area but with plenty of peril in between, or has time only for an abrupt upbringing of the offspring so as not to miss the migration, one wonders if the powers of natural selection are not being rather too devious for our comprehension. There is even entirely contrary behaviour within the same species. Some individuals remain in the same place without undue suffering; others fly off and experience considerable hardship. Those that remain frequently survive in greater numbers.

Part of human incomprehension must be an inability to absorb the innumerable factors that are involved. Just because migration is one word, there is no need for the event to have one cause, one reason, one answer. Migrations, says Jean Dorst of Paris, are like the birds themselves; they are multiple, and involve different elements which cannot be reduced to a rigid formula. There may be a reason. There may have been a reason. There may be a variety of reasons, of which the most important – in our eyes – may be the least valuable. Many migrations, according to us, and with our existing knowledge of the situation, should be cancelled. They seem foolhardy, unwise, pointless or suicidal.

The reason why swallows fly south poses sufficient problems in itself, but the when, and how, of migration are just as intriguing. What is the urge? Why does a caged greylag goose, entirely separate from its fellows, suddenly modify a peaceful acceptance of the captive situation into a frantic desire for escape? And why does it wish to leave in one direction only? Any answer is immediately confounded by the activities of some other migratory species. What is good for the swallow, which may always follow a water course, is not good for the gander which soars high over the Tien Shan, the Takla Makan desert, the uplands of Tibet and the formidable fortress of the Himalayas. What is applicable to a bird is nonsense to a butterfly. What makes sense for a migrating flier, in that it can generally see horizons, does nothing down below for a migrating swimmer whose visual range is far less embracing. Explain why a martin returns to the same eave and one still wonders how the Alaskan fur seal finds again the small speck of the Pribilof Islands. Consider it reasonable that warmer, brighter days bring on the breeding season and its migration, and then be amazed at that New Hebridean bat, which lives all day in a dark unchanging cave, which flies at night into a climate noted for its constancy, and yet manages to become pregnant every year always at the same time. Why, and how? Each question is sufficiently complex on its own. Yet the actuality of the events follows a smooth and effortless course. Here is an example.

The greater shearwaters *(Puffinus gravis)* fly in a large circle every year. They start at Tristan da Cunha, the volcanic intrusion into the South Atlantic where these birds breed from January to March. Suddenly they leave, fly north, cross the tropics rapidly, and congregate off Newfoundland at the start of June, having accomplished some 6,000 miles (assuming only a direct course) in the intervening weeks. Still travelling they move north-east to the vicinity of such places as Greenland, the Faroes, Iceland and Britain. By August they go south again, but with less haste and in a less uniform manner than on their northward course. Some straggle down the Old World's coastline, some even go straight back via Newfoundland and the way they came, while others take a more mid-Atlantic route. In any case, barring those lost on the journey, the returning birds head, purposefully or casually, for Tristan da Cunha where the cycle will start all over again. The shearwaters treat the vastness of the Atlantic Ocean much as farmyard ducks treat their own round pond, from latitude 37° S to 60° N and then down to 37° S again, a minimum oval trip of 15,000 miles necessitating two crossings of the equator. For the remainder of the year they fly about Tristan, and rear another brood of tireless, trans-Atlantic fliers.

Behind that story, and quite apart from the enormity of its scale, lie all the problems. What induces the birds to fly north in March? What is the precise trigger and what the general impetus, two distinct facets of the same act? What causes the splitting up in the North Atlantic? What again is the trigger and the impetus for moving south? How do the birds find the speck of Tristan da Cunha, no easy landmark for any navigator? And what has made the shearwaters ready for the breeding season when they arrive? The reproductive system cannot start to mature only after arrival. Similarly it cannot be too advanced before the flight has been accomplished. The migrational trek alone is astounding; its mechanisms are doubly bewildering.

Moreover any thoughts about mechanisms, and all possible conjecture about a bird's abilities to find food and to navigate successfully, can be thrown into disarray by the realization that many insects also undertake considerable migrations. An insect's brain, being such a modest collection of nerve cells, is scarcely worthy of the term, and certainly incomparable to the bird brain. Therefore an insect's migration appears even more perplexing. Take the case of the Monarch butterfly *(Danaus plexippus)*.

In the summertime these large brown butterflies are scattered all over the United States and southern Canada. In September they start moving south, at first independently, then with others, finally in huge congregations. They fly to Florida, to the land border of the Gulf of Mexico, to Mexico itself, and to California, with a preference always for warm places near the sea. Hanging on to trees they then live throughout the winter months in a dormant state. Insects cannot be said to hibernate, for that word is generally restricted to warm-blooded animals which become substantially colder.

Towards the beginning of April the long dormancy comes to an end, the butterflies leave their trees, and a northward trek begins. As with the start of the southward movement, the butterflies behave independently and move off singly: but, unlike the autumn migration, the insects all remain independent in the spring,

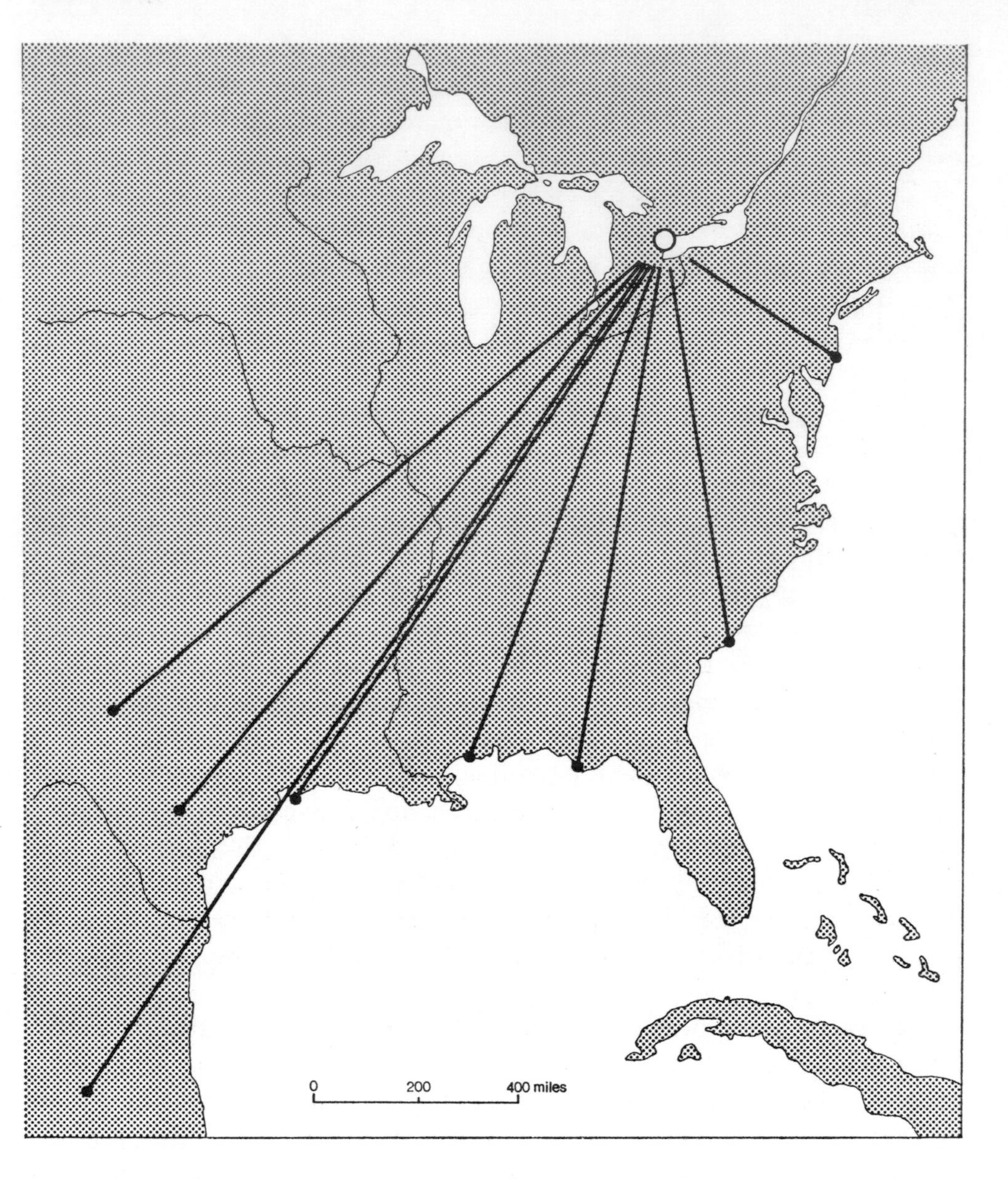

16. The lines on this map diverge from Toronto, the starting point from which eight Monarch butterflies were known to have migrated. The points indicated by solid dots show where they were recovered in the autumn of the same year. The 1,870 mile flight to Mexico was covered in 4 months, while one flight to Texas took only 42 days.

never forming large clusters. They fan out northwards and, as they do so, the females lay eggs on milkweed plants *(Asclepias)*. The trip north takes about two months, and the Canadian eggs are laid in June to hatch out either later in June or in July. More southerly hatchings have time to grow, to become sexually mature, to lay eggs of their own, and for these to hatch out, and even to mature and lay yet more eggs before autumn comes.

In the north, with a shorter summer, there is less time, and the single crop of butterflies is ready to move south at the end of August. All antecedents, whether they are the parents of the Canadian young, or the grandparents and great-grandparents of those born farther south, do not winter in the south again; they die. Therefore the descendants, whether first or even third generation offspring, migrate on their own. It is not the same with mammals and birds when the young so often travel with the old (although there are plenty of exceptions), and when it could be argued that the young are learning from the old in preparation for the time when they will be guiding yet further generations; the butterflies know the way without any such parental guidance.

The young Monarchs move southwards on their own, and head for those convenient trees in the coastal warmth of the southern latitudes. The whole trip from Canada to the south and back, always undertaken by inexperienced individuals, is over 2,000 miles long and is no less remarkable than the oceanic circumnavigation of the shearwaters. Yet from insect to bird the cause, trigger, impulse, navigational method and so on, can have few physiological parallels, if indeed any. The butterfly weighs less than half a gram, a seventieth of an ounce. It cannot be functioning in similar fashion to a bird and yet, somehow or other, it virtually repeats the performance. Monarchs marked and released in Ontario have hurtled off with bird-like determination. One Texan arrival had logged an average of thirty-two miles a day, and another covered 1,870 miles from Ontario (or more, as even butterflies cannot be expected to fly in a bee-line) before it settled down to spend the winter deep in Mexico. How on earth does 0·4 grams fly 1,870 miles in the right direction, rest and then fly back again depositing eggs correctly en route?

Nevertheless, both bird and butterfly seem to have sense to their wanderings. The insect advances with the warmth, and then retreats with it. The shearwater also follows the sensible practice of being in each hemisphere during its long days of summer, and the journeys are probably much influenced by food availability and basic wind direction.

The explosive migrations of the lemming appear far less reasonable. *(Lemmus lemmus* is the Norwegian lemming, the most famous in Western Europe, but the other lemming species, *Lemmus sibericus,* has a far bigger range throughout North America and the Soviet Union.) It is even argued that the lemming marches are not true migrations, in that a circular or return journey is not an intrinsic part of them; but there is nothing about the word migrant which impels all emigrants to return. However, the lemming story, a third example of migratory behaviour, is full of misconceptions, many of which have a devoted and ancient history.

The totally apocryphal lemming tale, incorporating all the traditional legends, would have us believe that, every so often, lemming armies come pouring from the Scandinavian hilltops, led by some fanatical master-lemming. They then consume crops, attack people, inject poison, foul all water, hurtle wildly over the countryside by night and by day, accept no obstacles as being too much for them, and finally seek some form of salvation in the ocean. Without hesitation, and frantic for self-destruction, they swim out and perish, every single one of them. Embellishments are that they are traditionally seeking out Atlantis, now submerged, or that the master-lemming, having achieved his genocidal whim, slinks back once again to his mountain home. All legends make one suspect that every story-teller is a journalist at heart. Also all the legends, as with journalism, have kernels of truth to them.

The lemming does indeed have outbursts from its normal concentrations in the southern mountains of Norway and Sweden, and from the low-level tundra farther north. The exodus is not a Gadarene impulse, but a collection of independent movements which add up to large numbers. There is no leader. There is no communal and mob-like movement. It is just that many lemmings are behaving in similar fashion at the same time. They travel mainly by night, just as they are normally nocturnal animals. They certainly are no danger to man, but they seem to instil fear much as a mouse can induce it because it is out of all proportion to their size. (A guinea-pig is roughly ten times heavier than an average lemming of 2½ ozs.)

The migrating animals have lost much of their own customary fear, but they do not plunge, irresolutely, blindly, madly, come what may. They take a way round, if there is one, and swim to the nearest promontory of the opposite bank. They are savagely pounced upon by dogs and all other predators on their journey. The losses are tremendous, and many do indeed drown, either in rivers, the lakes or in the sea; but there is not necessarily a 100% loss, and there is no Plaza Toro leader to return for a fresh consignment of victims. The slaughter is not absolute, but nearly so. The migration cannot be called suicidal, as if self-destruction was the sole aim. Instead – and there is a difference – a pattern of behaviour is induced in the lemmings which does lead to tremendous loss of life.

Recently the lemming seasons, the occasions of these migratory eruptions, have been occurring every fourth year, or thereabouts. Each exodus is caused primarily by overpopulation in their mountain community and is a symptom, as Charles

Elton puts it, of a maximum in numbers which is always terminated by a severe epidemic. It does not seem to be caused by a lack of food, or by some sudden preponderance of disease, and the strangest feature of this migration is the part (presumably) played by natural selection. How is this migratory urge perpetuated in the lemmings, since those that migrate generally die and those that do not migrate are the ones to multiply and replenish the Earth? One also wonders about the precise overcrowded moment when the migration is initiated. Who stays and who goes? If the urge to migrate is inherited the selective advantage conferred upon those lemmings which manage not to migrate, despite the headlong rush of the majority, must be colossal. And yet selection in this instance is an inadequate force; otherwise the migrations would have stopped long ago. The stay-at-homes would have bred other stay-at-homes, and the impulse to emigrate would have been snuffed out as surely as the emigrants; yet the lemming surges have been recorded for over 400 years, and are presumed to have a far more ancient history. In fact, to quote Elton again, the inherent tendency to migrate under certain circumstances appears to be so firmly ingrained that no ordinary selection in favour of its elimination makes any difference. It is part of the make-up of the lemming ... just as the lemming's lungs and brain are part of its fundamental make-up. Lemmings just do behave in this fashion.

The apparent or actual idiocy of the lemmings' migratory urge makes one take another look at the apparent sense of other migrations. It seems most reasonable to us if the movement is in time with the seasonal oscillations, and moves back and forth with an isotherm. It also seems reasonable for a species not to be static, to travel elsewhere – despite the losses, to invade different geographical areas, to be adaptable whether the grass is greener on the other side of the hill or not; but the lemming contradicts such reasonableness.

The lemmings are not alone in their perversity, only the most dramatic. All small and many large rodents appear to control their numbers by a regular succession of population peaks, of outbreaks, of sudden declines, and then of a slow climb back to yet another peak. Innumerable insects, such as locusts and many butterflies, have similar population explosions which may lead to devastation of the surrounding environment but which show slender comprehensible purpose. Human beings describe their own recent population growth as an explosion, although the global growth of 2% a year is hardly explosive in lemming and rodent terms, but the bursting of the bubble and the sudden decline of the population, as in the aftermath of every lemming year, is a distinct possibility for the human future. At the moment we can only try to accommodate our own numbers, and be distantly comforted by the regularity with which *Lemmus lemmus* does indeed burst its own population bubble, and yet thrives upon the determined departure of the bulk of its members.

Anyhow, faced by such migratory nonsense, by a blatant eclipsing of traditional selective pressures, it is easy to adhere to the second explanation of migrations. The first says there is a purpose to migrations, whether we can see it or not. The second suggests that migration is the result of an internal rhythm, a regular impulse to wander, to migrate, to get up and go. It is something harmonious, a fundamental rhythm of living things. It therefore needs no further explanation; it does not have to be advantageous; it overrides natural selection. It is a basic fluctuation, pulsing, ebbing and flowing, and only with the higher animals, with their increased mobility and more complex organization, does it manifest itself as something apparently needing additional explanation. This second theory suggests no difference between primitive meanderings of movement in a protozoan and migration in a vertebrate; both are rhythmic and basic, although one is more complex due solely to the more complex nature of the more advanced creature. Neither the lemming nor the single-celled animal, by this account, is its own master. Both are subject to the changing state within them.

It is always possible, in virtually any subject about which two extremely dissimilar explanatory theories have been put forward, that both may be right in different contexts. Some animals may migrate because it is advantageous for their species to do so; others may do so as a result of a fundamental impulse within them, and yet others may migrate both because it is advantageous and because of an inherent rhythm.

It would be hard for anyone to deny that two theories are likely to be involved, bearing in mind the enormous scale of the migratory impulse, and that there are daily, seasonal and irregular migrations to be accounted for. Dr C. B. Williams, the entomologist, has defined migration as a continuous movement in a more or less definite direction, in which both movement and direction are under the control of the animal concerned. He adds that there is often known to be a return movement to the original habitat, but he does not consider this to be an essential part of the definition. In fact migration is the rule rather than the exception and, chosen from a wide range, here are other examples.

Many whales (certainly the sperm and hump-back) breed in warm waters, and return to Antarctica to feed and grow. Caribou of northern Canada move south in July, reach the timberline, stay there, probably turn north again, and then go south in September to spend a wandering winter in the forests of midwestern Canada. Their young are born on the march in the spring migration northward. Europe is the scene of large bat migrations, and some have flown from East Germany to Lithuania (600 miles), a distance eclipsed by the Mexican free-tailed bat which annually flies from south of the border to New Mexico and Oklahoma. And there are all those larger mammals of the African game parks, like the wildebeest, which regularly migrate from controlled security into the well-poached lands over which no jurisdiction can protect them. Turtles are also great migrators. They hatch on the beaches, paddle out to sea, live elsewhere (but exactly where is frequently a mystery) and then return as adults to lay eggs on the original beaches. Some of their journeys are already known. Mid-Atlantic hatchings from near Ascension Island travel to South America, grow up and return to the Atlantis of their birth. On a different scale the molluscan chiton lives on coastal rocks, its small shell a vital protection. Every day it migrates away (or commutes) only to return to its precise niche on the rock. A Bermudan chiton, which travels about a yard from its selected niche every day, was once observed to have a far smaller mollusc on its back. This one too migrated regularly over the chiton's shell, always returning to the original position near a particular valve.

Most birds migrate, and it is highly improbable that any wild bird, particularly in the temperate world, lives its whole life within a few miles of the place where it hatched from an egg. (Incidentally humans are far less determined to move: 67% of those now living on the Isle of Man were born there.) The speed of bird migration is also fantastic. It has been proved by banding and subsequent recapture elsewhere that a blue-winged teal once flew 3,300 miles in twenty-seven days; a sandpiper flew 2,300 miles in twenty days; and a lesser yellow-legs flew 1,930 miles in six days. Distances are huge for the bird-size involved. The yellow-legs' speed of 322 miles a day was achieved by a bird weighing only 3½ ozs. The blackbird-sized Wilson's petrel nests south of South America, flies north in April, summers in the North Atlantic, and then hurries back for another nesting season. Not only is it a fairly small bird, but Mother Carey's chicken (another name for it) is a slow flier as birds go. Yet it may well visit both the South Orkneys and the Scottish Orkneys every year.

Not only can speed and distance be staggering, but height too. Much migration takes place at night at about 2,000 ft (the invention of radar was instrumental in proving this point) but there are remarkable exceptions. The high-flying geese have already been mentioned, but even rooks – which do not give the impression of being outstanding fliers – have been seen at 11,000 ft. where oxygen supply is two-thirds of the sea-level quantity. Curlews and cranes have been observed higher than 20,000 ft, and thus living on less than half the normal quantity of oxygen for each given volume – or lungful – of air. Geese have topped them all by flying higher than Everest where the air (as human climbers panting into their oxygen masks know only too well) is one-third of its sea-level density. Due to the phenomenal dimensions of bird migration, it is small wonder that we think first of birds as the migratory animals, but there are many others.

Most sea fish migrate, first to the spawning grounds, then to the feeding grounds.

Many river fish migrate seasonally up and down the rivers. Salmon and various others do both. They migrate down the river to feed in the sea, and then up the river to spawn. Eels are similar but contrary; they feed in the rivers and then spawn in the sea. Many insects migrate – butterflies, moths and others. For example, nine of the eighteen species of hawk moth normally to be found in Britain do not survive the winter, but fresh stocks migrate in from Europe every year.

Locusts have the most venerable history as migrators, and devastations are known to have plagued Africa and Asia for millennia. Confronted by a large swarm a man once calculated, on the assumption that he could slaughter a million locusts a minute, that it would have taken him seven days and nights to destroy the swarm. Apart from these huge irregular outbreaks there are also regular seasonal migrations. The desert locust of Africa produces one brood in the north in winter, and another in association with the southerly rains in summer. Many beetles migrate. Ladybird movements are particularly famous, both in Britain and in the United States, where the Convergent ladybird flies hundreds of miles every autumn to a winter hibernation, and then hundreds of miles back again to the orchards. Plankton migrate vertically in the sea, changing depth by night and by day. Crabs, like the edible *Cancer pagurus* found around Britain, move in September from the shallow tidal zone to deeper water of about 150 ft where they spawn.

In short, animal migration is by no means confined to the famous migrators. It is widely practised. It is often incomprehensible. It is a natural desire to wander, said Robert Collett. It is an animal characteristic of great antiquity. It is a part of animal life. However, it is frequently triggered by the seasons, and frequently the survival value of a migration is associated with the regular environmental changes in each year. But the seasons are not the entire answer, and it is this very multiplicity which is so confusing. The word migration is about as all-embracing as the word travel for human beings. Of course there is no single explanation or trigger for travel, and by no means is travel always advantageous or even sensible for humans, however basic the impulse.

Finally, and throwing a great cloud of ignorance over a subject already fairly obscure, there is the problem of navigation. How does a chiton find its niche? How does a young bird know where to go? (The young North American golden plovers leave for South America even before their parents, and always travel a different route.) How does a ladybird, which certainly does not have the eyesight or the mental equipment of a bird, succeed in emulating that bird? And how does another bird, living in the almost regular daylength, steady heat and evergreen environment of a tropical forest, appreciate that the tundra thousands of miles to the north will soon be losing its winter mantle of snow?

On the actual process of navigation Jean Dorst has written that to know precisely

Bird migration Birds can migrate for extraordinary distances in response to the changing seasons and may literally span the ends of the earth. The Arctic Tern (ABOVE) winters in the Arctic and spends summer in the Antarctic, a flight of over 10,000 miles.

BELOW: Swallows spend temperate summers in the British Isles and Europe, then return to the sub-tropical summer of Mozambique, Natal and the Cape. The reasons for flights of such immense distances and the navigational methods involved are far from clear.

RIGHT: Even the tiny humming-bird is capable of long and accurate migratory flights, covering up to five hundred miles at a stretch.
BELOW: The Junco bird normally commutes between Canada and the U.S.A. Experiments with artificial light by William Rowan in the 1920s first established the connection between increasing daylength and migration in this and other birds. Migration is tied up with factors such as breeding and moulting, the accumulation and consumption of fat, and may even change diurnal activity into nocturnal.

Animal migration Birds are not the only migrants.
ABOVE: The wildebeest in the African game-parks regularly migrates from controlled security into more dangerous territory. Many do not survive to make the return journey (RIGHT). In some cases it can be argued that migration has no definite purpose and is perhaps not much different from the primitive meanderings of a protozoan.

LEFT: Turtles are also great travellers: some are known to have hatched on Ascension Island in the mid-Atlantic, made the journey to the coast of South America, and returned as adults to lay their eggs in their place of birth.

LEFT: Lemmings, found in Norway, North America and Siberia, are renowned for their march to the sea. The millions that join the march mainly die. Migration seems to be a response to overcrowding in mountain areas rather than the mythical suicidal impulse or legendary search for Atlantis.

LEFT: Pilot whales washed ashore at Fort Pierce, Florida. Many whales, including the sperm and humpbacked types, breed in warm waters and return to the Antarctic to feed and grow. These whales died after resisting efforts to deflect them off course out to land, so strong is their directional instinct.

LEFT: The Monarch butterfly spends its winters in California, Florida and Mexico, and at the beginning of April moves north towards Canada to lay its eggs, covering distances of up to two thousand miles.

LEFT: Seasonal migrations and irregular plagues of locusts have troubled Africa and Asia for centuries. The African desert locust breeds twice a year, once in the north in winter and once in the south after the summer rains.

LEFT: Geese are known to reach heights of 30,000 feet crossing the Himalayas on their migratory flights. At such heights the oxygen supply is less than half that at sea-level, and the casualty rate is very high.

Hibernation The deep sleep of winter hibernation involves an extreme temperature drop in a normally warm-blooded animal (to around freezing point) with an accompanied reduction in energy needs. It is a kind of survival of the weakest, a total retreat from uncongenial conditions.
LEFT AND BOTTOM FAR LEFT: The dormouse accumulates mounds of winter fat and has a dramatic drop in heartbeat from the normal 250 beats a minute to twenty-five.
BOTTOM LEFT: The hedgehog's spikes are good protectors but bad insulators; it remains dormant for twenty-three weeks, during which there is no food intake.
TOP RIGHT: The ground squirrel, an American species, hibernates although its European counterpart does not. The sawdust on its back is to help researchers trace its every movement.
BOTTOM RIGHT: The only bird known to hibernate (and only two cases have been observed) is the American poorwill, a member of the nightjar family. The bird here was accidentally discovered in 1946 and spent three months in a hibernatory state.

ABOVE: Greater Horseshoe bats in a Dorset cave. Bats in cool temperate zones hibernate when necessary, hanging upside down.
RIGHT: Marmots, native to northern Europe, Asia and North America, live in colonies and prepare for hibernation by carrying bundles of hay into their burrows.
BELOW: The black bear is not a true hibernator, but goes into a winter torpor for up to six months. Many animals do this, including swifts, humming birds, queen wasps, mackerel and basking sharks.

what the individual bird does, to know how it takes its direction, how it compensates for the wind that bears it effortlessly like a swimmer in a strong current, how it reacts when confronted by fog or storm, what it does should it overshoot its mark – to know these things accurately one would have to *be* a bird. Perhaps, but that implies that living individuals know why they react and behave in a particular fashion. If a young man's thoughts in spring turn to those of love does he have any notion of the events which have hit him or of all that he does in response to them? One can presume that the bird is no less vague as it is impelled to go north, to travel, to arrive safely. Once again there must be a mixture of factors. There must be an impulse to travel, and there must somehow be guidance for the direction of travel.

This second point, exemplified by the ability of the shearwater to find Tristan and of the fur seal to find the Pribilofs, has caused the wildest theories to be promoted. Many of them have their origin in mankind's frequent inability to accept the fact that animal sense organs have a sensitivity of quite a different order to those of man. A dog has a nose about a million times more sensitive (for many odours) than the human nose. A hawk, hovering high above a field and looking for mice, plainly has exceptional eyesight. A bat's ears are incomparably efficient. The migrating animal, impelled by instincts, may be guided by factors which human beings know about, but cannot detect. Somehow or other those young, unescorted golden plovers of North America do head south and do reach South America, after a journey of 8,000 miles on their own. Something must guide them on their way, the midday sun, the spin of the Earth, the warmth ahead, the smells. They do not operate in a vacuum, any more than a newborn baby which fumbles about looking for a nipple. A baby does find it, with or without maternal assistance, and the birds do find South America.

Now to the seasonal triggers and to the actual physiological changes taking place in a regular and annual migration. The birds are not only the principal migrators in most people's minds, but they have also attracted most attention in scientific circles. Before mentioning a few of the scientists' results, and the implications, it is as well to describe a typical migration from the various viewpoints.

In the first place this generalized bird leaves its southern quarters, flies north, nests and then returns. Two triggers are therefore necessary for this simplified version, the trigger to go, and the trigger to return. However, to change the angle and to complicate the story, the bird must have stored fat before the journey, it must be in a receptive state for the triggers to operate, and the effect must be the correct one. (Shortening days in the southern hemisphere's autumn may cause a bird to fly north across the equator. Shortening days in the subsequent northern hemisphere's autumn may then be the trigger for the bird to fly south.) The bird must be sexually mature on arrival at the breeding ground. It must be psychologically reactive to any group or mob activity at the right time. It must rear its young, put on fat again, become receptive to the next trigger, fly south and not overshoot the mark, also at the right time. Finally, and looking at the migration from a more

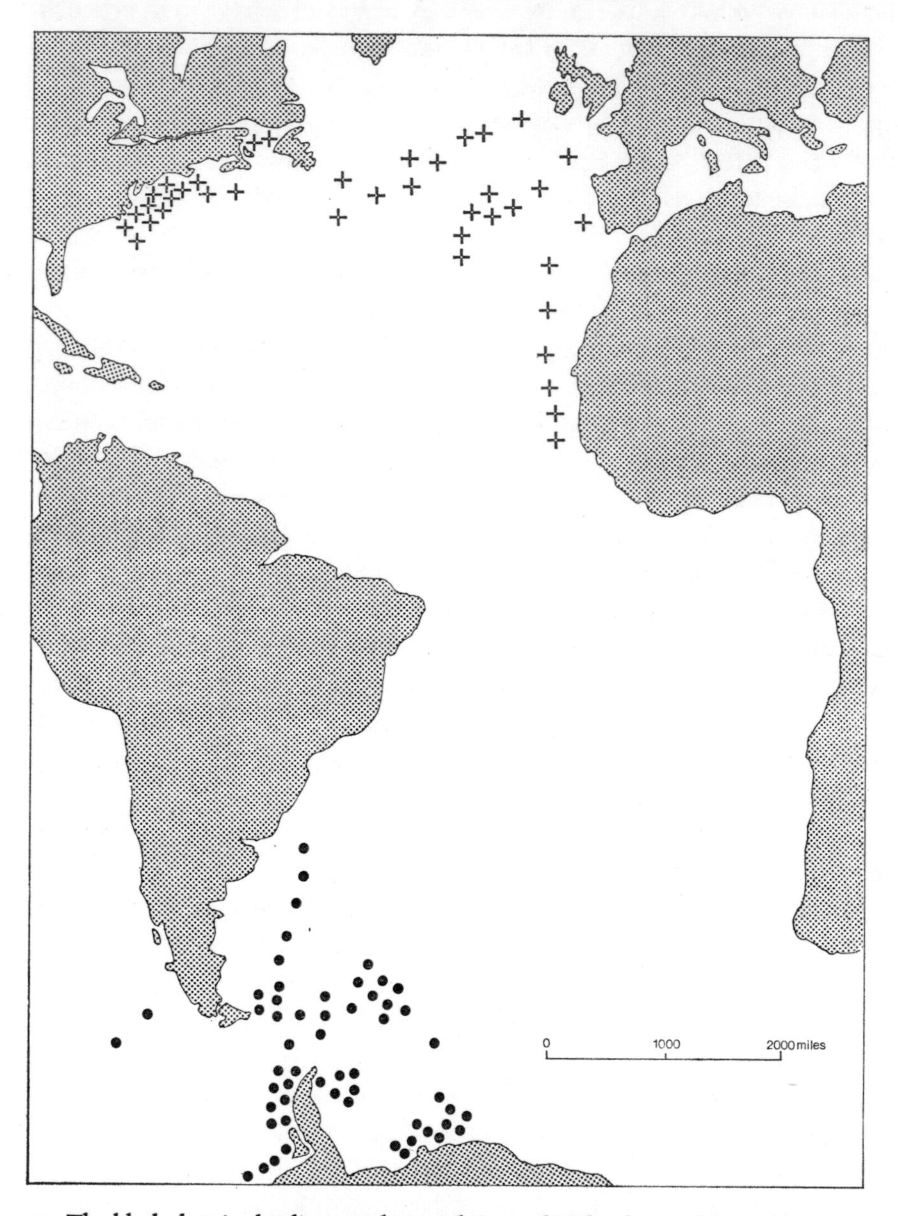

17. The black dots in the diagram show sightings of Wilson's petrels in January, at their nesting places off South America; crosses represent points in the North and South Atlantic where the birds have been sighted between June and October.

teleological point of view, the precise timing of the journey may be linked with, for example, the sudden emergence of suitable insects for feeding the young. This means that some southern trigger, such as changing daylength, must correctly initiate the migration so that the flight is accomplished, the nests are built, the eggs are laid and then hatched so that the new crop of gaping beaks does coincide precisely with the sudden abundance of those suitable insects.

Even if a fixed astronomical yardstick like daylength is the initiator there must be modifiers on route, speeding up or slowing down the process, so that there is coincident timing between a fledgling and insect. After all, a hotter, wetter or cloudier year can advance or retard those insects, and a lack of liaison leading to a lack of food for the young birds can be disastrous. Migration is not just a get up and go. It is a complex of interactions precisely as interrelated as any ecological situation involving several species.

Having said all that, what about the physiological basis of it all? A bird does not set off by magic, or magically lay eggs on arrival. There is a biochemical reason for the deposition of fat, for the enlargment of the gonads, for physiological change. It was William Rowan who opened up the subject in the 1920s and 1930s in his all-important experiments largely upon the junco *(Junco hyemalis)*. Normally this bird migrates north to spend the summer in Canada, and then returns to winter in the United States. Rowan suspected that lengthening days were the spur for going north, and vice versa. So, artificially, he subjected the juncos to different amounts of light per day. He discovered by this means that he could switch on their gonadal development, and then switch it off by subjecting the birds to shortening days. In fact he managed to do this several times a year, instead of the once induced naturally.

He also proved that temperature had no effect, but that exercise – caused by jolting the cage – had considerable effect in developing the gonads. (At once the plot, as they say, thickens. Is it longer days themselves which develop the gonads, or the increased activity which is commensurate with a longer day?) He also discovered that daylight can affect even blinded birds, proving that the optic nerve is not a necessary part of the system. Rowan's initial conclusion was that daylength formed the trigger, but an internal cycle predisposed the birds to that trigger.

As soon as Rowan's results were published, the acclamation – and the objections – poured in. What about tropical daylength, which is so constant? And what about birds arriving from the other side of the equator? For an answer he modified his general laws, notably by the addition of a refractory phase, a time after breeding during which triggers have no effect. If no cartridge is in the chamber, the trigger cannot be effective and nothing can be fired.

Further problems related to Rowan's initial experiments were encountered during subsequent work. If daylength increase causes gonads to grow, and if growing gonads cause migration to occur (a Rowan hypothesis) how could it be explained that many birds migrate even after castration? And how did it happen that some normal birds arrive at their breeding destination without any gonadal enlargement? It was soon realized that Rowan had discovered an association between daylength and migration, but to link all together all the time plainly had its pitfalls.

Other scientists did similar work upon the restlessness of caged birds. In general it was found that these birds did become more active at migration time, even without any of the external stimuli. Cold weather tended to increase their restlessness. It was also observed that restlessness caused birds to lose weight and, conversely for some birds, weight gain caused increased restlessness. Just as a human castrate often puts on weight, so it was felt that the cessation of a bird's reproductive activity after breeding effectively castrated the animal, leading to weight gain, leading to migration, thence to weight loss, and so on.

Theories cropped up no less frequently than new migratory facts. Friedrich Merkel in 1940 injected thyroid extracts, and then argued that the thyroid gland was the power behind migration. Indeed there were proven links between this gland's activity and many of the physiological states before, during and after migration, but once again a nice theory was confounded by the birds themselves. Some seemed to obey it. Others, by the inactivity of their thyroid at a time when it should have been most active, did nothing of the kind. In other words, as with the gonads, the thyroid gland is part of the system, but not the dictorial determinant.

So what about the pituitary, that other unseen governor of physiological activity? For one thing it controls the thyroid; for another it influences metabolism in general, and the reproductive cycle in particular. It can also act independently of outside influences, as can be shown by taking birds to the other hemisphere and watching them continue their old behaviour six months out of phase with the seasons. So the pituitary, influenced both by light and by its own internal rhythm, can be thought of as the *eminence grise* of migration; but is there an even more acute power behind its throne?

After all, what causes its internal rhythm? Scarcely anything is known about the reason or forces behind a gland's metronomic ability (but there is still the chapter on biological clocks to come). Sometimes, as with those birds across the equator, which persist in their old timing, the metronome wins. Sometimes, as experiments have shown which artificially caused gonads to shrink, fat to be lost, weight to be lost, and moult to happen seven times in a year, the pituitary's rhythm can be displaced by changing light. Albert Wolfson of North Western University (in the US), who carried out many of these experiments, concluded that there is no cause and effect between migration and gonad growth, but the two accompany each other because of the activities of the pituitary, the gland which acts both on them and on all the endocrine glands. It creates a physiological state in which the development of a certain organ is part of the whole.

It should be remembered that experiments often fail to get at the whole truth because normal behaviour is curtailed. In life a bird loses weight during migration because of the energy expended. Therefore, devoid of fat and short of weight, it may be physiologically more receptive to the next stage of the cycle. Also birds may be stimulated independently into migration, but then be influenced by the psychological mob effect as all the other birds fly alongside; this may be auditory or visual (and difficult to recapitulate in an experiment). The full-time occupation of nesting, incubation and rearing the young may be making the birds totally non-receptive to the next trigger, until all that parental activity has been firmly concluded. (An athlete only starts running if the gun fires when he is on his mark; he does not start running at every bang.)

For the birds the gradually increasing daylength may act on its own, but it may act because it means more hopping about by the bird, or a greater consumption of food in the time available, or a greater availability of food, or increased courtship time. By no means need light itself be the actual determinant. Certainly the pituitary is influenced by light itself, but the pituitary is also influenced by the general metabolic state and activity, and these are both influenced by longer days. A scientist often longs for facts to come in neat packages. The endocrine control of migration does not appear to be such a package.

To sum up, and riding rough-shod over the exceptions, changing daylength is of prime importance. The pituitary is the main effector, being itself affected both by external stimuli and by its own internal rhythm. The simplicity of the migration act is complicated by associated activity, such as breeding and moult. The flight itself is associated with other changes, such as the accumulation and consumption of fat, and often the substitution of nocturnal activity for diurnal. Some birds migrate because the cycle itself is strong enough to impel them; others have to wait for triggers. The effect of the pituitary, important to migration, is of tremendous importance to the body as a whole. By no means have its properties yet been evaluated in detail, either for migration in particular or for the body in general.

However it is still fantastic that small feathered objects, weighing half an ounce, fly 5,000 miles to Britain every year, arrive on a similar date, seek out the same area, produce their young and then all hurtle back over the 5,000 miles to Mozambique, Natal and the Cape. The annual migration of the swallow is an extraordinary response to the changing seasons. The actual sequence of endocrine events may yet turn out to be more remarkable than the flight itself.

15 Hibernation

What can be done in response to an uncongenial environment? There only are four possibilities. The first is to leave the place for good, to emigrate. The second is to go elsewhere but to return when conditions are better, to migrate. The third is to stick it out and then, if the uncongeniality remains sufficiently intense, to die. The fourth is to adapt, to find the situation less exacting by acquiring a different yardstick with which to measure it. Therefore it is in this fourth category that hibernation is placed. It is adaptation of a high order, a reduction in temperature inevitably partnered by a huge reduction in energy needs. Consequently, although the environment is savage, the hibernating animal lives off its fat and waits for better days.

The process of hibernation is undoubtedly a cunning adaptation to cold, but it could also be called a clear case of the survival of the weakest. Not only are hibernators those animals which can neither fend for themselves during the winter, nor store up enough food and fat to satisfy winter activity, but they are animals whose temperature regulation is bad. In general, and under favourable conditions, they perform poorly in limiting their heat loss. The hedgehog, for example, has good spines for protection, but they are bad for insulation. The ordinary house mouse is able to make use of the winter for it remains active and can rear families; the dormouse (not a mouse, but named after the French *dormeuse,* meaning sleeper) merely retreats into hibernation and rears no young ones. However, as compensation, hibernators do live longer than equivalent non-hibernatory species.

It may appear negative, but it is probably useful to start off by destroying a few misconceptions. Squirrels (of Europe), badgers and bears, for example, do not hibernate. True hibernators are very rare. Among the mammals only a few species of rodent, insectivore and bat do so. These groups are among the primitive rather than the advanced mammalian lines. Within the birds there is only one true hibernator, although swallows and various other migrants have a venerable reputation of doing so. Also hibernators do not pass the winter in a state of complete immobility; they frequently wake up. There is a similarity with sleep here, in that no one sleeps like a log and everyone turns over once every twenty minutes or so throughout the night. Hibernation is not merely a simple, prolonged sleep. There are entirely distinct physiological events during it, such as a much reduced heart beat and a lower temperature, which have no parallel in a night's sleep.

Many animals do enter a state of torpor, quite a different situation from sleep. Torpid examples are found among the mammals (like the brown bear), the birds (like the swift), and even some of the fish (like the mackerel and the basking shark), but such torpor is not hibernation. Also cold-blooded animals, however recumbent they may be in winter (like queen wasps), cannot be called true hibernators.

So what is hibernation? L. T. Pengelley and Kenneth C. Fisher have defined it as a situation involving animals which can maintain a high and constant body temperature, usually close to 98·5° F (37° C), against the normal range of environmental temperatures, but at certain times under natural conditions abandon this homeothermic state and permit the body temperature to fall close to, if not quite to, the environmental level, with a low limit of about 32° F (0° C), from which they are able to rejoin the homeothermic condition at any time against the environmental gradient. In other words, the animal must be efficiently warm-blooded to start with, it must permit its temperature to drop to around freezing point, and it must be able to return to its original warm-bloodedness even though the existing outside temperature may still be cold. Put another way round it means that warm-blooded animals acquire some of the characteristics – within certain limits – of cold-blooded animals, and then revert to normal.

By the Pengelley/Fisher definition no invertebrate or fish can hibernate. So the word torpor is applied to them as well as to certain warm-blooded animals. Undeniably something has happened to the metabolism of all the animals in a torpid state, and the phenomenon deserves a name even though it is not true hibernation.

In many ways, as will be seen, migration and hibernation have parallels. Both are answers to the problem of winter. Both are triggered by external events (hibernators fail to hibernate if they are moved to warm spots) and both are affected by an internal or endogenous rhythm acting on its own. When warm laboratory hedgehogs do not experience either temperature or light variation they do not hibernate, but they do gain weight every winter and lose it before every summer, indicating that the old rhythm is still present. However, whereas migration is a frequent occurrence in the animal kingdom, with many creatures being migrators, hibernation is not. In most warm-blooded animals the heat regulation system works too well for it to fail on demand, and the heart and the nervous system do not function correctly (in man and many other warm animals) when body temperature falls.

The striking achievement of the hibernators is not merely that a controlled drop in temperature is permitted to happen, but that the various organs still function after the drop in temperature. The heart beats very much more slowly and does not begin to fibrillate (or experience tremor and irregularity) as it does when most warm animals are cooled. For instance, the human heart fibrillates at 64° F (18° C) or thereabouts, and the human nervous system stops transmitting impulses at 50° F (10° C). Therefore the human being is not a hibernator because even the sense organs of hibernators continue to function despite a temperature of perhaps 36° F (2° C). The animals can be awakened from their frozen state by noise or light (bats are particularly alert hibernators despite being at a temperature that would have killed all humans).

Today's hibernators are probably those animals which were originally from the tropics, and which have always performed poorly in regulating heat loss. On being subsequently surrounded by a harsher environment (following a move away from the tropics) they did not greatly improve their regulation mechanism, but did succeed in evading the problems of winter by transforming this poor heat conservation from a deficiency into a benefit. Man is also a tropical animal, having a poor mechanism for maintaining heat (shivering and such hair as he does possess are poor substitutes for a decent fur), and so he might have been a suitable hibernator. Unfortunately man, and all the other higher mammals, as well as all the largest mammals, cannot suffer the necessary drop in temperature and live. Nevertheless there is speculative appeal in the idea, particularly for men in the colder parts of the temperate world, and particularly if six months' annual dormancy could step up the life span to seven score years or more.

So who does hibernate? Certainly the hedgehog, which is Europe's only insectivore to do so. Not only is it a poor heat conserver, but its normal invertebrate food, such as insects and snails, disappear in winter. Certain *Scimidea* hibernate, such as the American woodchucks and ground squirrels. (The point should perhaps be made again, due to such a firm entrenchment of opinion to the contrary, that the squirrels of Europe, including the imported grey squirrel, do not hibernate. Indeed they frequently breed in the winter, thus indicating a degree of activity totally inconsistent with the deep, nearly frozen rigidity of hibernation.) Some of the *Zapodidae* (a mouse group) and of the *Cricetidae* (hamsters) do hibernate, but by no means all. Many bats do so but, reasonably enough, none of the bats which live in the tropics. And there is a solitary bird, the poor-will night-jar, *Phalaeonoptilus nuttalli*. More will be said about many of these examples later but, for the time being, the main point is the extreme shortness of the hibernation list. For the main part it comprises just a few cold-temperate rodents and some insectivores.

The pseudo-hibernators are far more widespread. These are the animals which spend time in a state of torpor and therefore do not fall within the closely defined aegis of true hibernation. Torpor is an unsatisfactory state in that it is hard to define, and the definers themselves find it hard to agree. A woolly generalization is that animals in torpor have lower temperatures than normal, but are still considerably warmer than the true hibernators. This implies there is only a matter of degree (or, to make the implied rejoinder, of degrees of temperature) between the torpid animal and the hibernator. Everyone is agreed this is an oversimplification of the case, and that significant physiological events, quite apart from a physical cooling, distinguish the two situations.

Unfortunately these events are not clear cut. There is a blending, as is so frequently the case in biology, between one group and the next. By how much must a metabolic rate be lowered before it constitutes true torpor. And by how much more before this torpor becomes genuine hibernation? As a further generalization, torpor is less of everything than hibernation. Temperature drop is less, heart-beat is nearer normal, inactivity is less pronounced, duration is less (many animals exhibit daily torpor) and arousal from this state requires less of a stimulus.

What animals demonstrate torpor? Many invertebrates survive the winter in an extremely stagnant condition with an exceptionally low metabolic rate, and invertebrate physiology does indeed have parallels with vertebrates, but the word torpor is sufficiently diffuse without using it to cover both the dormant insect and the dormant bird or mammal. Some authors use the word hibernation to embrace every form of wintry lethargy, but this renders it almost meaningless. By the same token, and to retain some form of identity for torpor, it is hereafter used for the vertebrates alone.

Fish are primitive vertebrates and some of them show torpor. It always used to be thought that mackerel migrated because they were only a summertime fish in European waters. Now they have been regularly trawled up in winter from deeper water, and it is there that they spend the cold, non-plankton season in a markedly reduced state of activity. The basking shark is another example, and will be mentioned again later in greater detail. Many reptiles become torpid, but they too are cold-blooded and so their torpidity is both harder to elicit and to define. Amphibia are similar. Among the birds both swifts and hummingbirds are known to become torpid frequently. Both have very stringent needs for a continuing bodily sustenance, and their periodic states of torpor lower temporarily both their metabolism and their demanding needs.

Among mammals many of the smaller creatures have daily torpor patterns. These periods of energy conservation and inactivity come between the periods of peak activity and food collection. It was the German scientist, M. Eisentrant, who coined the term 'lower warmbloods' to describe animals which have a marked daily temperature change, and this category includes those animals whose lower temperature limit plunges sufficiently deep to cause them to enter a state of torpor. However, there are also some mammals whose normal heat regulation is sound, but who can enter a state of torpor for the winter months; certain bears are an example. Their temperature does indeed drop, their heart rates slacken, their metabolism falls, and they are able not to urinate for several months. Despite all this the bears are not true hibernators, mainly because the temperature drop is just a few degrees. Hence the earlier conclusion that primitive mammals not possessing a good thermo-regulation system make the best hibernators. The more advanced mammals, like the bears, cannot evade their acquired thermo-regulatory efficiency, and consequently they consume far more energy in winter, however dormant they appear to be, and however inactive. Torpor is the poor relation of true hibernation.

Now to some positive hibernatory examples, and some details about their warm-blooded summers and cold-blooded winters. The European hedgehog *(Erinaceus europaeus)*, poorly protected from the cold, and always with a low skin temperature – a frequent attribute of hibernators – tends to retire into its cold and comatose condition relatively late in the year, probably during December. Cold weather is the essential spur. In the laboratory the hedgehog's hibernation can readily be initiated a month or two earlier if the animal is placed in a dark room at 39° F (4° C). Provided that food and water have both been withheld on the day before the cold treatment begins, the hedgehog will then hibernate within a few days, and will continue to do so for about 165 days. Throughout this time, if the conditions are sufficiently cold, the hedgehog's own temperature will be at 43° F (6° C). Obviously no animal can be colder than the surrounding medium, and so the external temperature will have to be lower than 43° F (6° C) for the hedgehog to be as cold as this, but the animal's temperature never falls below 43° F (6° C) – except in death – however cold the outside world. Should the outside world grow particularly bitter this can have the contrary result, not of cooling the animal further, but of waking it up, and thereby raising its temperature.

Although a hedgehog's normal temperature is about 93° F (34° C), and therefore its body cools by 50° F (28° C), the heat regulation system is still functioning, and is still, so to speak, aware of the situation. When the hedgehog reaches 43° F (6° C) it stays there. (Oddly this temperature is roughly the border-line heat between growth and non-growth in the botanical world.) The heart-beat of an active summer hedgehog is from 128–210 beats per minute. In the hibernating animal this falls to 2–12 per minute, and metabolism and energy requirements have, by then, dropped to about one-hundredth of their normal level. There is no intake of food during the hibernation, and so there is a weight loss throughout the twenty-three weeks of dormancy. As a percentage of the total body weight the loss is initially high. It falls to a minimum in January, and thereafter rises again. (The marmot also does not eat during hibernation, but can actually gain weight on occasion during its bouts of cold inactivity. The solution to this enigma is that, in the transformation of stored fat into carbohydrate, inspired oxygen is necessary and enough of this oxygen is chemically fixed within the marmot for a weight increase to result.) Despite the apparent intensity or depth of hibernation, the hedgehog arouses itself every eight to twelve days. Metabolism then increases, warmth returns, heart-beat and respiration rate both go up. After a few hours of near-normality everything then subsides again into the near-frozen somnolence of hibernation.

Nathaniel Kleitman, the American pioneer of modern sleep studies, has said that the hedgehog may be considered cold-blooded when it can economize on fuel by being so, but not when its life is endangered. It will cool off under cold conditions like any cold-blooded animal, but not below a certain temperature. It will then wake up. Mankind, by comparison, cannot even become cold-blooded, however desirable this might be from time to time. Nevertheless, man's temperature is certainly not a constant 98·4° F (37° C). Many people, particularly the elderly when sleeping in freezing rooms, cool a few degrees below that temperature, perhaps to 90°F (32° C), perhaps even less. Normally such a person will feel cold and will shiver. Shivering stops below 85° F (29° C), but blood pressure is then falling, the heart-beat slows down, and the victim feels depressed. The point of no return is about 75° F (24° C), for below that the heat temperature falls fairly rapidly and circulation fails at 70° F (21° C). One long standing human record for cold endurance is of a snow-drifted woman whose body was at 64·4° F (18° C) when they brought her in. She survived, miraculously, with no more injury than the loss of a few fingers and toes. However, the hedgehog, well bundled up, with nose almost to tail, and with all four limbs cosily touching both nose and tail, drops its temperature 21° F (11° C) more than that frozen girl, and suffers no injury whatsoever. Indeed it has everything to gain, losing no more than a lot of accumulated fat, and wakes up to encounter spring – and a fresh abundance of agreeable insects, slugs, snails and the like.

These intermediate arousal times have evoked considerable interest among researchers. Does the internal clock, or whatever system it may be, cause arousal after definite periods or at definite times? The dormouse *(Eliomys quercinus)*, that bundle of fur frequently found in its wintry nest in Europe, and from which it can be taken and handled, has proved particularly valuable for studies of the basic circadian – or daily – rhythm in hibernation. Normally this creature is nocturnal, becoming active at sunset, with the rule being broken only during the intense activity of the breeding season of May and June. Hibernation usually lasts – in Western Europe – from October to March, and the animal's temperature then oscillates in tune with the ambient temperature, keeping about 1°F (·5° C) above it but never going below freezing point. Similarly its heart-beat drops from a normal 450 to 25 beats per minute.

It is during this hibernating time that arousals happen. These periods of active wakefulness occur most frequently at the beginning and the end of winter, and least frequently during January when the animal may be inactive for twenty days at a stretch. The duration of each period of activity is always more than four and a half hours, always less than ten. The point in question in this instance is whether, despite the depths of hibernation, the animal still arouses itself at sunset just as it used to arouse itself after each day of sleep.

In France, M. C. Saint Girons, National Museum of Natural History, Brunay, S. et. O., kept some dormice for three winters at a constant temperature of 41° F (5° C) and in boxes shielded from daylight. They still aroused themselves, apparently quite normally. (Some control dormice were, of course, part of the experiment.) The intriguing result was that the encased and thermostatic dormice did wake up at sunset. Moreover they woke earlier both as the year advanced and as the sun itself set earlier. In early October they woke (or aroused themselves) on the average at 7.30 in the evening. By mid-November it was 6 p.m. By mid-December (the solstice) it was 3 p.m. They were then returning to their nests long before midnight, as against staying up till 2 a.m. in October. Saint Girons' dormice showed some departure from this pattern in the slightly longer days of January, but times of arousal did, in general, get later. By the last week in February the animals were waking at 5 p.m. and returning to their nests for a further bout of hibernation at 1 a.m. On no occasion throughout the entire winter did a dormouse witness sunrise, and it was noted that there appears to be less of a correlation between returning to the nest and sunrise than between sunset and leaving the nest.

Clearly no visual clue to the setting sun, or even the low and misty midday sun of winter, could have acted as an arousing stimulus for those dormice in their heat-regulated and lightless boxes. The stimulus to wake up must have come from within, and remarkably effectively, for it was still timed to the passing days even after a twenty day period in the seemingly unconscious state of hibernation. During arousal there were only a few hours of activity when it was possible (at least with the animal normally active) to check or re-set the mechanism for the next time. Therefore hibernation may continue for, say, 480 hours, then wakefulness for six, then hibernation again for, say, 560 hours, then wakefulness for seven, and every time the waking up occurs within an hour or two at most of the actual sunset taking place beyond the darkened box. It is a remarkable persistence of the basic rhythm. Imagine any human shut up in a dark room, quite so accurately knowing the time after twenty days. (There is an account of the people who have spent long periods of time underground in the section on biological clocks.)

There must, one assumes, be a very important reason behind these hibernatory arousals. It is highly unlikely that they are just some manifestation of restlessness, for so much of the slender reserve of energy is consumed during these periods of activity. It has been calculated (for the American ground squirrel) that the arousals once every three weeks lead to nine-tenths of the total winter expenditure of energy although they last for only one-tenth of the time. This is another but more emphatic way of saying that the hibernating ground squirrel is only consuming about one-eightieth of the energy it would be using if it were active all the time. There must be a cogent reason for waking up from such economical behaviour, and then squandering limited energy resources so dramatically, even though the waking periods are short.

As another positive example of an animal which retreats from winter, and as a change from two definite hibernators, the bear well exemplifies the differences between winter torpor and winter hibernation. Outward signs between the bear in his den and the hedgehog and dormouse in their respective bundles suggest an identical reaction to the harshness of a cold environment. Some bears 'den up' for as long as six and a half months, their body temperatures do drop, they do not urinate for months at a time, they live off bodily reserves of fat, and a denning bear can be no more aggressive to an inquisitive human than a dormouse. Its torpor, in other words, can be very deep. It can also be very light, for many cubs are born in winter. Their mothers cannot be expected to be in deep torpor at the time, and many bears at least lift their heads as soon as someone peers into the cavern of their winter quarters. After disturbance an animal may even walk around a bit in a sleepy fashion, and then return to its special form of slumber with its head dropped between its paws.

Unlike the hedgehog, whose heart practically stops and beats at from 2% to 10% of its normal pace, the bear's heart-beat only drops to some four-fifths of its usual speed. A lethargic bear in summer also has a reduced heart-beat, and this characteristic is merely accentuated in winter. There is always a big discrepancy between the heart-beat of an active, or excited, bear and one in a comatose condition, whether in summer or in winter. (Human hearts can also pound, but there is less difference in pace between an active human and one who is asleep than there is between an active and inactive bear.) Bear body temperature is reduced in its torpid state, but only by about 9° F (5° C). It is this relative lack of temperature change which prevents the bears from being classified as hibernators. Humans are even less effective hibernators, and are normally only 1° F or 2° F (about 1° C) cooler in their sleep, however deep it may be. The bear is also different in possessing such a remarkable ability not to urinate during winter. Known facts are few, but one particular Grizzly was known not to have urinated between 5 January and

5 April one year, and it might not have urinated for two months beforehand. It did not even bother to do so for two days after waking up from the winter stupor.

Three adult black bears which had to be killed (for other reasons) provided interesting facts about their heart-beat under cold conditions when they were subjected to lethal cooling. The experiment was carried out by G. Edgar Folk, of Iowa University. The bears were first tranquillized, than anaesthetized, and then cooled for four hours. Their hearts stopped beating when their body temperatures had fallen to 70° F (21° C), 63° F (17° C) and 61° F (16° C) respectively. Such temperatures are not particularly cold, but with two of the bears the heart-beat had fallen to one to four beats per minute at the time of heart failure. With dogs, for example, which have no torpid abilities, and which have also been cooled in this manner, their hearts have always stopped working before their heart-beat has been less than eighteen beats per minute. In other words, despite the bear's failure to be considered a true hibernator, it does show the hibernator's characteristic of maintaining a slow heart-beat in a cooled heart.

Nevertheless the bear, so frequently thought to be equivalent in its winter behaviour to the hedgehog or dormouse, is nothing of the kind. Another animal, the basking shark, is never assumed to have affinities with either dormouse or hedgehog, but in fact this fish has just as much right to have them as the bear. The basking shark does enter a winter torpor. It does behave in very similar fashion to the bear.

It was in 1953 that a Dutch paper was published (from A. B. van Deinse and M. J. Adriani) that described the discovery in winter in northern temperate waters of basking sharks *(Catorhinus maximus)* without gill rakers. These organs had always been assumed to be an indispensable possession of the plankton-feeding basking sharks, as indispensable as teeth to lions, for the rakers sifted the vital plankton from the water. Therefore, or so it was thought, if these sharks were without rakers they could not eat.

Unfortunately this deduction was confounded at the same time by the discovery of much plankton in the stomachs of rakerless sharks, although that discovery is, in a sense, irrelevant to this particular story. In any case the plankton may have been caught when the rakers were still in position because the slowing down of metabolism in winter may have caused a considerable slowing down of the digestion process, so that the plankton may well have been gathered long beforehand. The significance of the story is the discovery that some sharks spend the winter in British waters where, with or without rakers, they could not possibly get enough to eat.

A paper published in 1954 by H. W. Parker (of London's British Museum of

Natural History) and M. Doeseman (of Leiden's equivalent establishment) pointed out that a feeding shark travels at two knots (which is about as fast as an average human swimmer), and that the power required to achieve this speed in a twenty-two foot shark with a mouth open twenty-eight and a half inches, presenting a frontal area of 4·3 sq. feet, is one-third of a horse-power. This is equivalent to 212 Calories per hour (or 212 kilocalories). The shark's tail and body are inevitably an imperfect propulsion mechanism, and the real power output is nearer 265 Calories. Also, as there is inefficiency in any organ's conversion system, and a discrepancy is bound to exist between energy actually required and energy which has to be taken in, the calorific value of the plankton consumed has to be far higher than 265 per hour to meet this need. The estimated requirement is 663 Calories per hour just to swim along with an open mouth; the question is whether the open mouth would actually receive enough food to make up these calories. A further assumption is that the same shark should be able to filter about 16,000 cubic feet of water an hour. But in winter, when plankton are far less abundant, this volume of water would contain only 2·75 lbs of plankton, and they would possess only 410 Calories. To expend 663 Calories in order to gain 410 is plainly uneconomic.

It always used to be thought that during winter the shark must bask in waters with sufficient plankton to supply its three to four ton bulk with the necessary energy, a place where the shark would not lose on the deal as it cruised along. Based upon the discovery of the rakerless sharks, and upon others caught since in wintry and relatively plankton-free waters (such as the Irish Sea), the present theory states that basking sharks always shed their rakers late in the year, and they then descend into deep water, probably just off the continental shelf, where they become torpid and reduce their metabolic needs. In consequence, they survive. Probably the large quantity of oil stored in the fish's huge liver (although there is much dispute on this point) is the source of most of its energy needs, as it lies, inertly, many fathoms deep, waiting for spring, for the stirring up of essential minerals, for the sudden flowering of plankton as new warmth and new minerals have their combined effect. New rakers will have grown by then, and the enormous animal will emerge from its torpor to eat again.

A basking shark's needs are prodigious, but even less demanding fish like mackerel experience a similar life cycle. They too feed on the plankton, and they too used to be victims of a convenient theory which explained their wintry disappearance by suggesting that they migrated north every autumn. In fact they stay roughly where they have spent the summer, but their migration is downwards. This was discovered when bigger and better post-war trawls started bringing up mackerel at a time when, according to traditional thinking, they should have been well to the north. It is now known that they spend a quiet winter near the surface of the continental shelf. They return to the surface of the sea as soon as the plankton has conveniently reappeared.

Both shark and mackerel are cold-blooded; hence the inability to argue that they come within the narrow definition of the true hibernator. Birds and mammals are the only warm-blooded creatures, and it used to be thought that only certain mammals could experience the cold sleep of the hibernator. For one thing, or so the point was customarily made, birds were much more capable of moving swiftly elsewhere than the ground-based mammals, and therefore had less need to stick it out for the winter. Then, in 1946, this idea was shattered by an American naturalist in the Chuckawalla mountains of the Colorado desert. Edmund C. Jaeger found a night-jar, called locally a poor-will or goatsucker, lying in the hollow of a granite rock. It was torpid and still there ten days later. Its temperature was a remarkable 66° F (19° C) (as against its normal 106° F (41° C)) and both its breathing and heart-rate were much reduced. Jaeger weighed it, put a ring on a leg, and observed the bird for eighty-eight days. Then it disappeared with the coming of spring (the original discovery having been made in December). For the next four winters it returned to the same niche, and spent about three months in this genuine state of hibernation. So far, it is the only proven avian species of hibernator and, so far, only one other poor-will has been found hibernating. Even so it turned many traditional tables upside-down.

Other birds are alleged to hibernate, but they only go part of the way and for far less time. Hummingbirds, for example, are of quite a different kind of dimension to the bulky basking sharks, in that the sharks are half a million times more massive than the birds, but the birds also suffer from an over-demanding metabolism requiring persistent nutrition. The resting metabolism of the hummingbird is proportionately higher than that recorded for any other animal, and even this is increased sixfold when the bird becomes active. The speed of the wing-beat, the perpetual jerking from flower to flower, and the fighting or the aggressive demonstrations which are so much a feature of hummingbird existence – all these make it difficult to believe that the bird is acquiring sufficient sustenance even when it is actually feeding, let alone when it is temporarily away from its source of nectar. Not only is its metabolism reduced to a sixth of the active rate when it is resting but, at night, the rate goes down still further. Its metabolism during this nocturnal torpidity is one-twelfth as active even as the daytime resting rate, and the birds are then cold, incapable of flight and without movement. This seemingly lifeless state, as it was first described by Professor Hans Krieg of Munich, is short in duration and cannot be called hibernation. Many scientists at the time of the original announcement (1940) did not call it anything. They sceptically dismissed the findings and found it impossible to believe Krieg's story.

Whereas hummingbirds feed in the daytime, and drop their temperatures mainly at night, nestling swifts are able to behave in a similar fashion whenever food is in short supply. An early fable, widely believed until the late eighteenth century, stated that many migratory birds, and swallows in particular, formed clusters in the autumn and then dived en masse to the river bed to pass the winter. The story probably helped to harden scientific opinion against any hibernatory theory for these birds, and cold-blooded torpidity was equally hard to accept. Nevertheless Krieg's thesis did receive support. Dr David Lack of Oxford demonstrated that, when bad weather or other conditions prevented adult swifts from catching sufficient insects, the brood in the nest adapted themselves to the situation by dropping their temperature and, in consequence, their needs. The nestlings could therefore withstand a period or periods of fasting which would have led to the deaths of other similarly sized birds. An intriguing addendum to this tale is that, for ordinary taxonomic reasons quite unrelated to these recent discoveries, swifts and hummingbirds and night-jars had all been classified in closely related families. The subsequent revelations have happily emphasized their relationship.

So much for the hibernatory and torpid activities of animals in general and a few animals in particular. Once again, having outlined a piece of behaviour, there is the need to account for such behaviour. Some mechanism or mechanisms must account for reproductive activity in early summer, for the deprivation of fats later, for the onset of hibernation, for the periods of arousal, and finally – perhaps the most baffling of all – for the awakening in spring that leads, once again, to the onset of reproductive activity. Hibernation has so many parallels with migration. Both events are seen to happen, and with both there is purpose in their happening; yet the causes of both, the precise workings of the clockwork, are far more elusive.

In 1964 R. A. Hoffman summed up the situation in mammals that hibernate by recognizing four definite phases in this annual cycle of events. First is a warm-blooded phase. It is spring. Sperm and ova develop, although testes reduce their weight. The thyroid gland is active until mating takes place. Then both the thyroid and the male reproductive system degenerate. It is during this time of sexual anticipation and fulfilment that hibernation cannot be made to happen, however powerful the stimulus, however intense the cold. Hibernatory animals at this time respond to cold as ordinary mammals always do. They step up their metabolism in order to keep warm and, should such counter-measures prove inadequate, they will die.

The second phase, during which the young are born and cared for, is a time of preparation. Reproductive organs and the thyroid are relatively quiescent, although the testes may suddenly become active again just before hibernation, but food consumption is high and much fat is laid down. Any external cold at this time stimulates the preparation for hibernation: metabolism goes up, more food is consumed and the time for entering hibernation is brought nearer. At first sight it is an anomaly that the higher the metabolic rate the more rapidly the winter state is hurried along, but obviously the faster the preparation the sooner the animal will be fit to enter hibernation. At the end of phase two any cold conditions, whether artificial in the laboratory or genuine outside, will cause hibernation. At this time the animals are ready to behave quite unlike the great bulk of warmblooded creatures. All preparations have been completed.

Hibernation itself is the third phase. The animal is dormant, but the reproductive system and various glands of the endocrine system gradually become active. The hibernating animal has to be prepared for its intensive springtime activity after waking up, and consequently the ductless glands have, so to speak, to prepare the preparations. Just as the sperm have to be ready before copulation in spring, the reproductive system has to be ready to mature the sperm, and the endocrines have to stimulate the reproductive system. Everything has to anticipate everything else. Anyway, phase four, the final phase, is the greatest enigma. The animal finally does wake up – somehow. Cold weather brought on hibernation but what stops it? At present no one knows but, despite ignorance on this score, the fulfilment of phase four is most emphatically the start of phase one, and yet another turn of the cycle.

Amplifying Hoffman's condensed four-phase picture of the hibernator's annual chain of events are various germane extras. For example, injections of the male sex hormone testosterone will prevent the onset of hibernation. (In other words a phase one situation injected into phase two will prevent phase three.) Injections of the hormone thyroxin at the same time will have a similar effect. If the adrenal cortex – yet another of the endocrine glands – is totally removed from the rat, a non-hibernator, and the animal is then subjected to cold, it will itself soon grow cold and die. If the cortex of both its adrenal glands is removed from a hamster, a natural hibernator, and if that animal is then placed in a very cold environment, it will behave more like a normal mammal confronted by a very cold climate. It will not hibernate but will try to keep warm, and will die after about six weeks. If the same hamster is given deoxycorticosterone, an adrenal cortex hormone, some of its hibernating ability will be returned to it. In other words a temperature regulating system in a normal mammal is seen to be a hibernation regulator in a hibernator.

The endocrines are more crucial than cold weather to hibernation in that no amount of cold will induce hibernation if the endocrine pattern is wrong, and cold is quite unnecessary for hibernation if the endocrine picture is as it should be at the end of phase two. (Hormone injections can even lead to hibernation in high summer.) Cold conditions in winter, although important in initiating hibernation, are far less important in calling a halt to it: hibernation continues for its allotted span even if temperature and light and daylength are all maintained artificially longer at midwinter levels. There is plainly an internal cycle at work which depends partly for its timing upon external factors and partly upon internal physiological events. In no circumstance is this more apparent than in the event of waking up from the winter sleep.

A clue lies in the nervous system. A human being's nervous conductivity stops when body temperature falls below 70° F (21° C), but hibernating animals 35° F (19° C) cooler still possess a lethargic, yet effective, nervous system. Noise or gentle touch will wake them. Their electrical brain rhythms can be detected during hibernation, and the proof and end result of all this nervous activity is that the animal does not permit its temperature to drop below a certain level. To make that point is easy. For the animal to carry it out, to increase respiration, to consume more food, to increase blood flow – all this involves considerable metabolic adjustment. Whatever the crucial difference is between a hibernator and an ordinary mammal it must be partly a basic function, possibly at the cellular level, of nervous tissue.

However that does not explain the mechanism of arousing spontaneously, as against being aroused, save that a functioning nervous system is plainly more capable of being an alarm clock than is a non-functioning system. It has been suggested that the accumulating presence of certain metabolites, which have been created by such metabolism as there is during hibernation, would be increasing irritants to hibernation. Unfortunately for this theory, urination does occur during the spontaneous arousals, and the theory loses ground when it has to postulate instead that certain irritating metabolites are not excreted, for some obscure reason, during these periods. The great unknown of hibernation arousal will have to wait for the future before it is clarified.

As a final point the study of hibernation is by no means an esoteric subject, entirely distinct from human needs. Not only are the physiological changes likely to enlarge knowledge about warm-bloodedness in general, but it would be highly advantageous if certain men, notably space men, could hibernate. What is the point of travelling even to the nearest other solar system only to grow senile in the process? Moreover, why grow senile at all? The study of hibernation could well elucidate a few points on that score. Whether immediately relevant to humanity or not, hibernation is a fascinating temperature response of a few warm-blooded creatures to the cold of winter and the lack of food in that season. It is eminently sensible as a system and, on occasion, undeniably enviable in its dormant unconsciousness.

A perplexing postcript to the whole subject of hibernation is that something similar happens in environments quite the reverse of the cold situation. Aestivation is the name given to the state of torpor experienced by some animals in extremely hot conditions. The perplexing point, of course, is that an escape reaction suitable for extreme cold is also suitable for extreme heat, but there is a reason, nonetheless, for the similarity.

Take the Mohave ground squirrel *(Citellus mohavensis)*. This is a rodent, active in the daytime but only during spring and early summer. In the winter it hibernates, and in the high summer it aestivates. The hibernation part of the annual cycle is not outstandingly different from that of other hibernators, save that this particular squirrel can, on occasion, enter the state of hibernation directly from its previous state of aestivation. This can happen in the autumn as intense heat suddenly becomes intense cold. The animal's normal, active and daytime metabolic rate is 0·8 ml. of oxygen per gram body weight per hour. This falls to between 0·1 and 0·2 ml. when the animals are aestivating. Their temperature at this time may have dropped by some 50–55° F (30° C) from the normal 97° F (36° C). Although entry into the torpid state happens rapidly there is no sudden temperature at which aestivation can be said to begin: at 90° F (32° C) the animal is still normal, but when its temperature is 81° F (27° C) it is probably aestivating. Even though the squirrel is then so torpid, with its breathing pattern being very different from that of a sleeping squirrel, and with occasional periods when breathing ceases altogether, the animal can be awakened very easily. It certainly cannot be handled and still remain inert as with most hibernating animals. It will, for instance, turn over if placed on its back.

However, its body temperature will obligingly fall if the ambient temperature falls and, should summer be merging into winter, aestivation will then become hibernation. Below 59° F (15° C), the squirrels will remain on their backs if turned over, and will become increasingly comatose and unwilling to respond as the temperature drops still further. Just as there was no hard and fast line between activity and aestivation, save that the metabolic activity was gradually reduced, there is certainly no hard and fast line between aestivation and hibernation.

Although aestivation is a word of considerable antiquity, the subject was never studied properly until 1955. In that year, C. P. Lyman and P. O. Chatfield of Harvard Medical School, made the bold statement that no body temperature of an animal in the state of aestivation had ever been measured. Shortly afterwards F. Petter of France made the first measurements and found that the rectal temperature of a gerbil *(Gerbillus gerbillus)* fell from 97° F (36° C) to 71° F (22° C) in a couple of weeks after it had become lethargic in August. Since then measurements have been made on many other mammals accustomed to severe heat for at least one period of the year, and many of them do indeed lower their temperature and their metabolic activity at the hot time of the year. Examples are the little pocket mouse *(Perognathus longimembris)* and the kangaroo mouse *(Microdipodops pallidus)*.

So why should a small mammal living in the heat of a North American desert react to its environment in the same fashion as another mammal living in the cold of a temperate winter? The answer is that a lowered metabolic activity does have advantages in both circumstances. The wintry hedgehog does not have to starve due to the sudden absence of its traditional foods, and the desert mouse does not have to starve because of the equally withering disappearance of its normal vegetative foods.

However, there is a further point about aestivation of particular advantage to the desert mice. As anyone knows who has taken the trouble to breathe on a piece of glass cooler than his body temperature, water vapour from his breath will condense on that glass. The same will not happen if a pair of bellows is used to blow ordinary air on the glass. In other words, inspired air is not only warmed in the lungs but picks up moisture, which is lost from the animal in expired air.

To a desert animal the loss of water is yet more crucial than the absence of food, and it will lose water whenever its expired air is at a greater temperature than the inspired air. This will happen even if the air being breathed in is already 100% saturated with water vapour because air can and does pick up more water vapour if it is heated. Therefore the breathing-in of air with a relative humidity of 100% followed by the breathing-out of air with a relative humidity of 100% is causing a severe drain upon the animal's water resources. A cold-blooded animal does not suffer this inconvenience because its body temperature drops as the ambient temperature falls, and its reptilian breath would therefore not be visible on a nearby pane of glass should it decide to blow on it. The aestivating rodent, although warm-blooded, and experiencing the advantages normally enjoyed by a warm-blooded animal, emulates its cold-blooded neighbours when it is convenient to do so. By permitting its temperature to fall it has the double blessing of a lowered requirement for food and a smaller loss of water every time it breathes.

Many aestivators also wisely block up their burrows during the hot time of the year. This means that the summertime fug they create becomes saturated with water vapour and water is not lost through the burrow's exit. The desert soil all around may be dry with typical summertime dryness, but the aestivators' cramped quarters will be conveniently snug with water vapour.

There does not seem to be any desert creature with the means of storing water (although some plants can do so). Animals have done the next best thing and made certain that their consumption of it is reduced to a minimum. In this context, despite the superficial similarity of a hibernator and an aestivator, there is dissimilarity: the hibernator is conserving food requirements and the aestivator is conserving water at a time when water needs dominate all other considerations.

16 Internal Clocks

Time and time again on the previous pages reference has been made to internal rhythms. Time and again this has been pinpointed as a basic attribute of living matter, both fundamental to the organism and crucial to its welfare. The word trigger, again used frequently, is an apt term, for it implies correctly that something exists to be triggered off, to be fired, to be stimulated. In most cases it is the basic internal rhythm which sees to it that there is something to trigger, that the time is ripe, that a particular organ will be receptive to the firing. A human being's clock becomes supremely evident after he flies to another continent, then yawns through lunch, sleeps through tea, wakes up at 2 a.m. and cannot then re-arouse himself at 8 a.m. An animal's internal clock is similarly demonstrated when, for example, the dormouse arouses itself from the depths of a wintry hibernation every three weeks or so but always at a time near sunset. A plant's clock proves itself when its petals open an hour before dawn, even when all hints of sunrise have been excluded. These internal rhythms are indeed basic to living matter, but they are at present far from being understood, though their existence was first noted in fairly remote times. Scientific evaluation of the phenomena is very recent, and it is significant that the order of interest has gone from plants to animals and finally to man.

The name of Alexander the Great is infrequently found in the scientific literature, but it was on Alexander's march to India that Androsthenes, a Macedonian philosopher, noted that there is a daily leaf movement among the *Papilionaceae*. It is not recorded whether he also realized that such timekeeping also occurred if the plants were kept in the dark, but certainly experiments of this nature were being carried out in the scientific upsurge of the eighteenth century. Later still it was discovered that the characteristic was inherited. Inevitably there were sceptics who could not believe, despite the evidence, that the astronomical event of daylength had actually imprinted itself upon the behaviour pattern passed from one plant generation to the next via the seeds, but gradually the idea gained ground, and support. 'The periodicity . . . is to a certain extent inherited,' wrote Charles Darwin in 1880.

Animal rhythms were not observed until later, partly, one assumes, because an animal's daily routine is more complex than that of a plant; but gradually animal evidence did accumulate. Bees have a time sense in that they arrive at a plant community during the period of the day when it has most nectar. Insects frequently become active only at particular hours, and will continue to do so when in the laboratory and when denied all external stimuli. Denial of stimuli has always been a difficulty in these experiments, for not all the stimuli are realized as such beforehand. It used to be thought that infra-red light was unimportant to plants. It is now known that the cumulative effect of two minutes of light per day as a door opens and closes cannot be discounted, or even the rhythm of the scientists themselves, coming and going at certain hours. The current tendency is to assume that anything might be a time-setter, a hint for the adjustment of the basic rhythm. Therefore every possible factor, however remote the possibility, should be excluded from interfering with the experiment just in case it is relevant.

There are plenty of examples of clocks at work. Some tropical biting insects only bite at certain hours of the day, thus leaving the victim free for the next voracious bunch. Fiddler crabs organize their lives in keeping with the rhythm of the tides, and will continue to do so when locked far away from the sea. Many fish have a daily rhythm; so do birds and mammals, and so do animals right at the other end of the spectrum, such as algae. The single-celled, primitive forms responsible for so much of the sea's phosphorescence in the tropics have a twenty-four hour rhythm to their luminosity. They are bright for twelve hours, and then relatively subdued for the remaining twelve hours of the day. At first it was presumed that such an action was a reaction to the greater light of the sun, or to some other external stimulus – such as temperature – oscillating in tune with the daily cycle of events, but hundreds of generations of these marine algae have been kept in constant darkness and they have still held to their original schedule of twelve hours bright, twelve hours dim. The periodicity is inherited, it would seem, to a considerable extent.

Mankind's rhythms were the last to arouse scientific interest, but these exist as positively as with algae or the higher plants. There is certainly not just one basic rhythm, but innumerable rhythms, each pulsing away in its own particular fashion. Take sleep, for example. A baby is born with a rhythm of sleeping and waking, and one has to assume (until brain-wave recordings are successfully made on foetuses to prove this point) that this has been established long before birth. The timing of the on-off sleep pattern then lengthens with increasing age, until such occasion – one of relief for the parents – when there is only a single phase of sleeping and waking in every twenty-four hours, but the pattern is still rhythmic. So is body temperature, which fluctuates regularly over 2–3° F (1·5° C) in every day, being lowest in the small hours and greatest in the latter half of the day.

There are also rhythms to the excretion of urine, to pulse rate, to blood pressure, to the intestines, to the uptake of oxygen in the lungs. Just as anatomical efforts were made to find the seat of the soul in the human frame (Descartes believed it was in the pineal body), so have efforts been made to find out where the ticking has been coming from for these human rhythms. But the more this subject has been taken apart the more evidences of rhythms have been unearthed. As if some desperate man were trying to find the sparkle in a diamond, the more he divides the original stone the more its sparkle glitters back at him. Intestinal cells, for example, have been grown on their own by ordinary tissue culture methods. With careful examination it can be seen that they too follow the same rhythm as the intestinal cells still forming part of the whole organization. The rhythm, or sparkle, still exists, however small the fragment, however single-celled.

The existence of a rhythm, however varied, is plainly of importance if events are to be anticipated. (A watch not only tells you that you have arrived on time, but when to set out so that you will arrive on time.) It is obviously satisfactory not to wait until the sun is shining before a plant opens its leaves, but to have them ready and waiting for the sun. It is important for bees to make journeys likely to be fruitful, when the nectar will be present in the group of flowers for which they are aiming. A sexual season must be started in anticipation of the warm well-lit days of spring. The birds must leave for their migration in order to arrive at a suitable time. These rhythms have to be accurate to be of value. Dr R. B. Withrow has calculated that the clock's precision must be within 1–3% of the real time if a flower is to achieve its actual date of flowering within a particular week. Spring and autumn are the two seasons of the year when the length of the day is changing most rapidly, but even at the period of maximum change – the equinoxes – there is only fourteen minutes difference in length of day from the start to the finish of the equinoctial week at a latitude of 30°. Therefore the clock has to be extremely accurate to distinguish between a period of darkness lasting twelve hours seven minutes and a period of darkness one week later which lasts for eleven hours fifty-three minutes. Nearer the equator the weekly differential is even less, but it is greater on each polar side of latitude 30°. Therefore clocks regulated by change in daylength can be less efficient with increase of latitude, but a certain degree of accuracy is essential if some event, such as flowering or breeding, is to be initiated at a particularly advantageous moment in the year.

One further important example of the human and animal need for an awareness of time will demonstrate why the term clock is jointly valid for both the mechanical and the biological kinds. It concerns the problems of navigation which can be overcome by making use of the various astronomical aids, principally the sun and the stars. The finding of latitude, by measuring the height of the sun at midday, has always involved a fairly simple procedure for human navigators. It is only necessary to determine the maximum height of the sun above the horizon during each day, because the angle of that maximum point will give the latitude, provided the date is known. The sun's maximum angle varies precisely at any point according to the time of year. Therefore, if a bird, for example, wishes to know its latitude, however subconscious the desire may be, it must know the time of year. And the more accurately the better.

Longitude was a greater problem for human navigators, because it involved knowing accurately the time of day rather than the time of year. Therefore a chronometer was necessary, a time-piece capable both of withstanding the buffeting of ocean travel and of keeping good time for those inevitably long periods away from any checking device. In 1714 the British Parliament offered the huge sum of £20,000 for such a device. Eventually, and in 1762, a chronometer developed by John Harrison passed the tests. It was his fourth attempt, it satisfied its critical Admiralty examiners, and it meant that longitude could then be assessed with far greater precision. During a five-month voyage in 1765 one of his marine timekeepers lost only fifteen seconds, and less than a second a week is a remarkable feat even today. The story did not end there; Harrison's claim for the prize money was initially brushed aside by Parliament, and only after an exceptionally long fight, which probably only triumphed because he had the king on his side, was he given his award.

There is pathos as well as triumph in Harrison's award; he died shortly after receiving it. There was also sadness in that no one could copy his clock or its accuracy at that time. The first accurate *and* reproducible clock was made years later by Thos. Earnshaw and John Arnold.

Animals cannot have Harrison chronometers but they have to possess some kind of time appreciation if they are to make use of heavenly bodies in their navigation. It is best if this appreciation can include both time of year and time of day. Migrating animals make use of a variety of aids to help them reach their destinations, but it has been proved that many use the sun and the stars as guides in good weather, and therefore a good clock is necessary to interpret the changing pattern of the sky. Humans paid £20,000 for theirs (and then a further award to Earnshaw and Arnold); the animals had to acquire theirs through evolutionary development. As a result, both animals and people are now able to set off and find oceanic specks like Tristan da Cunha and the Pribilol Islands.

The navigational digression emphasizes yet another point and yet another parallel between man-made time-machines and endogenous rhythms. Earlier chronometers had been useless because they required constant re-setting and could not maintain good timekeeping over a long period. Biological clocks also need constant re-setting and usually get it from the great astronomical events themselves, such as the daily routine of the sun and the monthly routine of the moon. Jürgen Aschoff of Germany, world famous expert in this field, coined a name for them: zeitgebers.

This word has been generally adopted for these master clocks which force the biological oscillations to run more true. Human existence is surrounded by zeitgebers, and a little thought about them will practically tell us the time, however short we ourselves may be of actual clocks; have the children come back from school, has the afternoon post come, has the paper been delivered and, of course, the biggest zeitgeber of them all, has the sun set?

An important experiment by Dr Klaus Hoffman of Germany, not only proved the point about clocks and navigation but showed, via a nice demonstration of faulty navigation, that the clock can be re-set. Some starlings kept out in the open had been trained to expect food in the most northerly of twelve boxes arranged around the circumference of their circular pen. The birds were then taken indoors where the light/dark periods had been adjusted. There was a six-hour differential between the internal day and the natural one existing outside, and noon inside was really 6 p.m. After sufficient acclimatization to this irregular indoor world the birds were returned to their outside cage. The sun, their zeitgeber, shone brightly in the sky and the birds looked for their food in the easterly box, not the northerly one. They had remembered their food training, they knew where the north food box was by the position of the sun in accordance with their internal clocks, but their clocks were wrong by six hours, by one quarter of a day. Therefore they went east, not north, to look for food and were 90° wrong, or one quarter of the sun's daily cycle. There is a game, set and match quality about this experiment.

The difficulty of planning experiments so that all possible zeitgebers have been removed has grown more difficult with the increasing awareness of an animal's or plant's ability to appreciate subtle cyclic differences in the day. To illustrate this point it is convenient to describe an outstanding experiment on the growth of the sprouting potato tuber. Blindly growing, or so one would have thought, the sprout might be expected to grow quite evenly throughout each twenty-four hours provided it grew within the homogeneity of a pitch black, thermostatic, and carefully controlled experimental chamber. The sprouting tuber plainly does not have the delicate sensitivity to outside stimuli that a good many other plants possess, let alone the animals, but some tubers were experimentally blanketed off from every imaginable fluctuation in their environment and tested. Temperature, darkness, noise, and the air itself were kept as constant as the experimenters could contrive. Then they measured the tubers' rate of oxygen consumption, an indirect indication of rate of growth. Under such even conditions it was expected

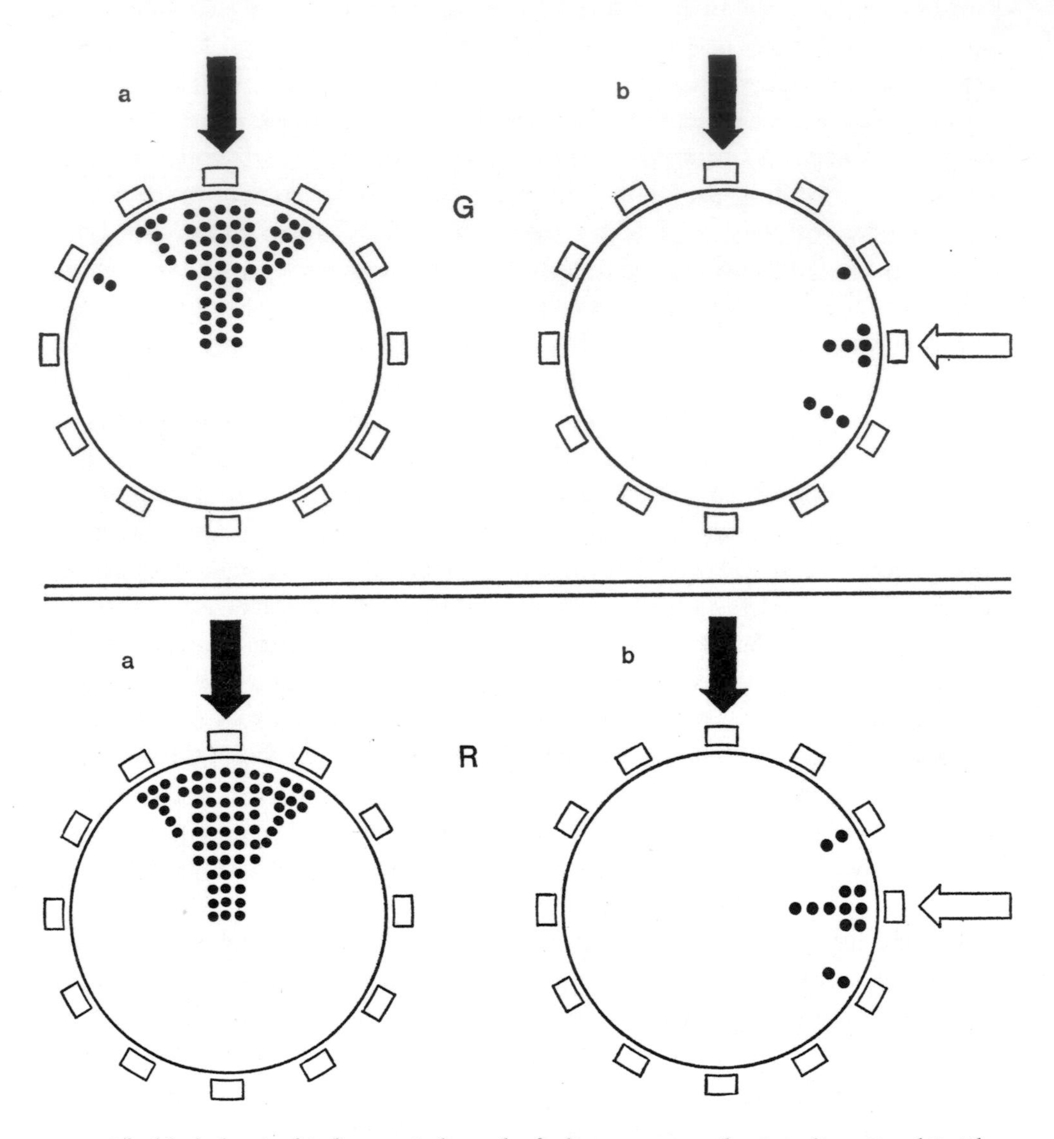

18. The black dots in this diagram indicate the feeding responses of two starlings G and R. The circular pens have feeding boxes ranged round their circumferences. The two pens marked (a) show the birds' response after being trained to feed at the most northerly box. The circles marked (b) show how the birds subsequently changed direction and took food from the box facing east despite their previous training and the presence of a bright sun to help them find North. This was due to the internal clocks of the birds having been affected by their exposure to artificially controlled alterations in the length of daylight. The navigational responses of the birds changed by 90° in response to their exposure to daylight periods artificially altered by six hours (or a quarter of a day).

to be regular. If found to be irregular, it was assumed that this would have no cyclic relationship with the hours of the day.

Amazingly the expectations and assumptions were proved wrong. The growing sprouts were both inconstant and rhythmic. Eventually, after trying to find all possible zeitgebers, it was realized that a statistically significant state of affairs existed correlating atmospheric pressure change between 2 a.m. and 6 a.m. every day and change in the potato's oxygen consumption on the same day between 5 a.m. and 7 a.m. As everyone knows, the atmospheric pressure is in a constant state of flux. Also, as most people do not know and cannot evaluate, there are some broad daily rhythms to these pressure changes, despite the irregular distortions to this broad pattern caused by the passage of the weather. The weather variations are considered to be random over long periods, but broad rhythms do exist. It was these which had a daily relationship with the well-regulated potato in the dark and with its rate of oxygen consumption.

Having appreciated this parallel, the experimenters then added a stable pressure to all the other constant factors of the darkened chamber. If pressure fluctuations were responsible for growth fluctuations, a constant pressure should eliminate them. However, and more confusing still, the parallel between the external atmosphere's daily rhythm and the potato's internal consumption of oxygen continued just as markedly even when the potato was kept in a constant pressure. This evenness, added to all the other constancies of its environment, merely highlighted the problem of running an experiment free of external stimuli. Even with such an experiment on such an object as the tuber's white sprouts, a zeitgeber existed which, in its own silent and undetectable action, told of the passing of the days.

Without zeitgebers, or at least without those which are important to each particular organism, the rhythm continues but continues erratically. A generalization is that the internal clock tends to slow down in the absence of zeitgebers. It is almost as if they help to wind up the system as well as to maintain its accuracy. At all events, the rhythm gets out of phase in their absence and slows down, perhaps by one hour in twenty-four. In fact, as Erwin Bünning of Germany pointed out, if an exact twenty-four hour rhythm is found under constant laboratory conditions, one should search for a controlling factor in the environment which has not yet been recognized. Just as no clock can be expected to show good time without occasional correction, so no diurnal rhythm can be continually perfect if denied its zeitgebers. If it does keep good time, a timekeeping zeitgeber must be presumed to exist.

The similarity between mechanical clock and biological clock can be strengthened still further. 'Oh that one's always slow,' says someone about his favourite clock, 'it loses five minutes a day.' Animals and plants can be equally consistent as their daily rhythm slips away from the traditional, well-maintained, twenty-four hour cycle. For example, a flying squirrel *(Glaucomys volans)* was once kept under constant and dark conditions. Normally its periods of activity started at about 6 p.m. in the evening. In the dark experiment this starting time slipped every day by twenty-one minutes plus or minus six minutes. In other words, its particular internal clock caused it to become active slightly later every day, never less than fifteen minutes later than on the previous day, and never more than twenty-seven minutes later. Such a consistent inconsistency is most remarkable. This regular aberration of twenty-one minutes is another way of saying that its own daily rhythm is twenty-four hours and twenty-one minutes but that, under normal circumstances, the external stimuli of an ordinary day shorten the time to a conformist twenty-four hours. The clock is slow but adjusted every day by about one minute an hour.

A particular bat, *Myotis lucifugus,* showed in a similar experiment that it had a basic rhythm of twenty-two hours twenty-five minutes in the summer but twenty-five hours in the winter. Once again there is a mechanical parallel. Some household clocks, particularly if they are kept in places susceptible to the warmth of summer and to the cold of winter, have vagaries according to the season. 'Oh, that one only gains in the summer,' they say. So too with *Myotis lucifugus.*

A parenthesis to this bat story has to remain somewhat muted because, so far as can be discovered, no evidence exists to substantiate it. However it is well understood that there exist two dominant kinds of human beings so far as rising in the morning is concerned; these have been called the owls and the larks. The owls are wide awake in the evening and they will happily stay up late, but they have to claw their way back into consciousness every morning from some nether region of sleep. The larks are those who wake up and get up, those who achieve a brisk walk before breakfast, and who then abjure any form of late hour. The theory, for which no substantiating evidence exists, holds that the human owls have endogenous rhythms similar to that *myotis* bat in winter, whereas the larks are the summer bats. In one the natural rhythm is never allowed, by force of clocks and life in general, to run its full twenty-five hour course. Jangling alarms always cut it short by one hour, and the system reasonably resents such unfulfilment. In the larks the system is alert long before any alarm, and long before the day is due. It is perpetually frustrated by the sun's lethargy in maintaining its dilatory twenty-four hour rhythm, when twenty-three hours are thought to be ample.

After all, to corroborate that point with argument if not with facts, it is unreasonable to expect that everyone's basic rhythm maintains a similar cycle. It is easy to imagine these rhythms being on either side of the mean twenty-four

Camouflage For many insects camouflage from predators is an essential response to environment. The leaf insect, found in the Far East and the islands of the Indian Ocean, merges into its feeding ground, never moving all day. Coloration is more complex in the case of reptiles and mammals where temperature regulation, courtship and even territorial claims are additional factors.

The white winter coats of some arctic and sub-arctic species are not directly caused by the onset of snow but serve to provide an effective camouflage.
TOP LEFT: A Snowshoe rabbit in winter. Compare the same rabbit in summer (BOTTOM LEFT).
ABOVE: The chameleon is well known for its changes of colour ranging from brown to bright green; this is the chameleon's only defence weapon, and occurs as a direct response to the immediate situation.
BELOW: The scorpion's main weapon is its poison, but it also relies on invisibility by night.

Rhythm in flowers ABOVE: The evening primrose opens at night.
TOP RIGHT: The gazania opens only in bright sunlight.
TOP AND BOTTOM LEFT: Dutch bulb growers use complex systems of temperature and light control to make their tulips flower as early as possible. The role of temperature in flowering is extremely complex.

BOTTOM LEFT: The camellia is one example of plants whose growth cycle can be altered by artificial hothouse conditions.
BELOW: Tobacco plants normally flower in the short days of winter but can be induced to blossom in summer if they are kept covered for the appropriate length of time each day.

Plant adaptation Plant species vary incredibly in their ability to cope with different environments.
LEFT: Poinsettia grows wild in hot places throughout the world and in Australia is regarded as a weed, but when treated with care it can be grown in temperate areas.
ABOVE: Cactus, found originally in the American desert, naturalises wherever conditions are suitable. The stem serves to store moisture.
BELOW: These mountain flowers on the slopes of Mount Kilimanjaro are found only above 11,000 feet.

hours. If this entirely plausible statement is true, it will explain the two sorts of waking human, with the larks being before the mean of twenty-four hours, and the owls being after it. Easy to propose, but hard to prove – save by experiment.

Human beings have in fact experimented on themselves in caves and potholes to test their reactions in timeless conditions. The general intention seems to have been one of record-breaking. Principal long-timers have been M. Siffre (1962: 62 days), G. Workman (1963: 105 days), T. Senni (1964: 125 days), D. Lafferty (1966: 127 days) and J. Mairetet (1966: 153 days). David Lafferty's four-month stay in a Cheddar cave yielded some remarkable results from that lonely world where he was given no hint from outside of the genuine passing of time and days. His day – that is the period from one awakening to the next – lasted from nineteen hours to fifty-five. His longest night was of thirty-seven hours, and a total of 60% of his time was spent in sleep, as against 33% for the eight hours a night of traditional behaviour. (One always suspects with human experiments that they cannot properly be compared with laboratory animal experiments because the potential boredom of the human being, when denied the normal interests of life, is likely to differ greatly from that of the mouse. It will either be greater or, if there is sufficient entertainment like books and music, it may be less.)

Despite Lafferty's early irregularity and his abnormal addiction to sleep he did settle down into more of a routine in the latter half of his self-imposed retirement, and days were often some twenty-four to twenty-seven hours long. At first sight of the results there does not seem to be much of a discernible rhythm in his sleeping and waking, but a complex mathematical analysis of his irregularity showed that there was an increased probability of his waking after intervals of twenty-five hours rather than after any other particular interval. In short, despite the vagaries, and despite the interactions of boredom and genuine fatigue and discomfort, there was a detectable rhythm, and this was slightly longer than daylength. His personal clock, as with those flying squirrels and with the winter bats, ran slow in the absence of its accustomed zeitgebers.

Michel Siffre, the Frenchman, who spent time near a subterranean glacier 375 ft below the Alpes Maritimes, was a much more regular case, waking later and later each day. Consequently his self-appointed time limit came to an end long before he felt it was due – 14 September, the appointed time, arrived when his personal and erroneous calendar read 20 August.

Anyone who misses a night's sleep knows that much of the ordeal is over by breakfast time on the following day. Even though it is then more than twenty-four hours since the last period of sleep, one generally feels more wakeful than a few hours earlier during the accustomed time for sleep. The pattern repeats itself should a second night's sleep have to be missed, or even more. Breakfast time is always a better and more wakeful period than beforehand. Modern living requires night-shifts, and the customary practice is for workers to have short spells of night-shift followed by longer spells of day-shift, possibly in the ratio 1:2. The dominant pattern is to work by day, and the daytime rhythms will persist with the night worker, causing body temperature, urine flow, blood pressure and other physiological patterns to fluctuate as if sleep were still occuring at night. Man is a daytime animal. It is abnormal for him to behave as if he were nocturnal and even more to switch, one week on, one fortnight off, from nocturnal to diurnal and back again.

It is also highly irregular for mankind to be hurtled, either sub- or supersonically, across the lines of longitude and to arrive, entirely out of phase, in another world. The airlines wish to have done with their passengers as quickly as possible and to evict them, overfed and underslept, to get about their foreign business. Consequently hunger, sleepiness, wakefulness, a desire for activity, and mental alertness among the transported passengers all come at the wrong times. The speed of travel is often immediately negated by the employer's insistence that no crucial business shall be transacted for at least forty-eight hours, but it takes far longer than that for endogenous rhythms to be reversed. Quite how long no one knows, but it could easily be discovered by dispatching people as quickly as possible halfway round the world. (New Zealand is excellently situated as an antipole for West Europeans.) Their external stimuli of all kinds, the sun and the daily living, would then have been shifted by half a day. It would be only too easy to measure the lack of shift of the various endogenous rhythmic features, or rather their gradual readjustment to this new hemisphere.

Even though so little work has been done on the subject, the few cave and pothole pioneers, plus various others living in the constancy of an Arctic day and a few more existing in the daylight-reversed world of a special laboratory, have already provided some interesting and basic leads. Not only are there the obvious rhythmic indicators of various organs, but the fundamental chemistry of metabolism has shown that it too fluctuates evenly. The excretion of sodium and potassium rises and falls throughout the day, according to an established pattern, and these maxima and minima are independent of working hours, resting periods, meal times or the type of meal. The maximum output is about midday, the minimum about midnight. When a man travels from Britain to New York both peaks remain at the same Greenwich time, but then begin their shift to US Eastern Standard Time. Sodium is the first to be adjusted, and potassium is some days later. It would seem therefore that these two elements have separate clocks. So too with other chemicals. Phosphate excretion is usually low at the moment of waking, and creatinine – found dominantly in muscle tissue and urine – is low at night.

Any attempt at asserting that such metabolic fluctuations are due either to internal or to external temperature changes should be resisted. Admittedly the temptation is strong. Every schoolboy with a test tube and every cook with a saucepan knows that chemistry happens faster with the application of heat. Anyone with a clinical thermometer can check that human temperature varies, and obviously physical work also leads to a rise in temperature, while cold nights might reasonably be expected to bring it down a bit, however effective the body's thermostatic mechanism. A rise of 18° F (10° C) roughly doubles the speed of a chemical reaction and the daily human changes could produce peaks, it might be argued, in the excretion of various electrolytes. However, a second thought will indicate that such a subjection to the dictatorial policy of temperature would lead to poor timekeeping.

It has already been mentioned that some mechanical clocks change their accuracy with the seasons, and the first biological experiments on the subject were dedicated to proving that the same situation must occur in life. Cyclic periods were expected to become shorter with warmer temperatures because warmer temperatures lead to faster chemistry and therefore less time for any given reaction. Oddly (and perhaps this proves some other point) the first experiments in the 1930s achieved the expected result. Both the leaf movements of certain beans and the time taken for fruit flies to emerge from their eggs were affected by temperature, in that the action was faster with greater heat. Later experiments, which exercised more control over some sources of error, led to entirely contrary results. The length of the various cycles (with bees being the principal creatures to prove the point) was found to be virtually independent of temperature.

In other words, because chemical reactions *do* work faster in the presence of heat, there must be compensating processes to keep the cycle more constant. It has been shown that the bee's life requires an effective clock for this insect to exploit the daily movement of the sun as an aid to navigation, and for the bee to know when to seek the nectar from a particular flower. Any endogenous rhythm that can resist the inevitable effects of temperature change, particularly in a cold-blooded animal so subject to external conditions, is of distinct selective advantage over any timing mechanism that changes, in effect, according to the direction of the wind, whether cold and easterly or warm and southerly.

Dr Brian Goodwin, the developmental biologist of Sussex University, has summed it all up by saying:

'The temperature-dependence of the free-running periods is actually much smaller than one would expect for a biochemical process, and this fact indicates that there is a temperature-compensation mechanism built into the clocks. If organisms, especially unicellulars, plants and cold-blooded animals, are to have reliable clocks which run fairly accurately throughout the year, then they must be relatively insensitive to seasonal change such as temperature.

In fact many organisms do even better. Not only do they compensate efficiently for temperature change but they use temperature change as another zeitgeber, or time-setter. The world is generally colder at night, and coldest in the latter part of the night. Therefore a regularity about temperature minima and maxima exists which can act as a zeitgeber just as much as any other rhythmic feature of the environment. Experiments have proved the point. The leaf movements of the plant *Phaseolus* have been controlled (to some extent) by temperature change. So has the sporing rhythm of the alga *Oedogonium*, the petal movements of *Kalanchoe*, and the general activity of the cockroach *Leucophaea*. Plainly such an ability is wide-ranging, and the normal laws of biochemistry are not only put aside but the temperature shift is insolently exploited as timekeeper.

Be that as it may, light/dark changes are generally more effective than cold/heat changes as clock-setters, and extreme cold will stop the clock. Some bees have given proof of this under experimental conditions designed to show that the cold actually stops their timekeeping systems, rather than the processes controlled by the clocks. Some bees, accustomed to warmth, were then kept for five and three-quarter hours at 39–41° F (4–5° C). Having been subjected to the cold temperatures they were then observed, and some controls were also observed, to determine how many bees, and at what hour, searched for food. The control bees were active, in the main, either before 10 a.m. or shortly afterwards (and most had given up the search by 1 p.m.). The chilled bees, even though warmed out from their earlier chilling, were less uniform in their behaviour, and few started searching for food until after noon. Their peak time was between 2.30 p.m. and 3.30 p.m., and searching was over by 4.30 p.m. Therefore the effect of the cold treatment was both to make the search behaviour less certain and to postpone the peak by some four and a half hours. The postponement was not quite as long as the period of cooling, but very nearly so. The clock, in short, had been stopped virtually for the length of the cold period.

Different organisms, of course, behave differently and clocks are stopped by dissimilar temperatures. A cooling to 41° F (5° C) was sufficient to derange the bees. Beans will be stopped if the temperature is lowered only to 50° F (10° C). The spider *Arctosa perita* has its time-sense put out of joint by a drop to 41° F (5° C). The pond skater *Velia currens* stops skating (and all other movement) when the temperature falls to 34–36° F (1–2° C), but its internal clock is still functioning. Cockroaches lose their sense of time around 50° F (10° C), but then they like warm conditions. The alga *Oedogonium* is active in cooler conditions and its clock only stops near freezing point.

A similar picture exists at the other end of the temperature scale when too much heat will also derange the clock. For instance, *Phaseolus multiflorus* is a plant with distinctive daily leaf movements. At 68° F (20° C) and under normal circumstances all the plants move their leaves. At 59° F or 77° F (15° C or 25° C) the proportion is nearer 60%. At either 50° F or 86° F (10° C or 30° C) only a quarter or less of the plants still possess their daily leaf movement, and at either lower or higher temperatures none do so. Once again it should be pointed out that these experiments have been devised to test the effect of temperature on the diurnal rhythms, not on the organism's ability to be active at certain temperatures. They have proved that modest afflictions of temperature do affect the clocks quite independently of their effect upon metabolic activity as a whole. Those bees were still alive. It was their clocks that had been stopped.

It has already been mentioned that efforts have been made to find the clock's epicentre, precisely where the ticks come from, as if there were some controlling agency working away and an object could be discovered like the instrument in Captain Hook's crocodile. This is where the mechanical clock analogy, which has held good until this point, breaks down. Any determined efforts to find the tick of an ordinary time-piece will most abruptly stop it. Like trying to find the purr of a cat, the tick will disappear. Not so the biological timekeeper. It is not only a feature of the whole organism but, apparently, of its component parts as well. In fact there is evidence that there is still ticking even at the cellular level.

Single cells from organisms are not the same as unicellular organisms, but the fact that single-celled animals have demonstrated that they possess endogenous rhythms suggests that such an attribute could also be part of the individual cells of larger forms. *Euglena,* for example, has provided evidence of its unicellular and diurnal rhythms. So has the protozoan *Paramecium* in its sexual activity, and so have various small but multicellular organisms, such as certain primitive algae and fungi. Their cells are outstandingly uniform, and the organism as a whole possesses an endogenous rhythm. The assumption is that the rhythm is probably a feature of each cell as well as of the cells as a whole. Therefore it should come as less of a surprise that certain tissues cultured from higher organisms have shown a diurnal cycle, that fragments of intestine half a centimetre long have shown a daily rhythm with their continuing cell division in particular, and their movement and activity in general. The tick of the clock, it would seem, is everywhere in general and nowhere in particular. It is, to make the point again, a basic attribute of living matter. Or so it would seem.

Now to the most awkward timekeeping question of them all. On what system or systems do these biological clocks operate? There are theories, needless to say, but that is all. There is no real knowledge about the mechanism. A huge stumbling block was ably shown by those blind tuber sprouts, growing unevenly in their even environment. No one has yet experimented on a terrestrial organism removed from this planet, from all possible influences of our terrestrial field. A satellite in orbit (and that includes the moon) is still blatantly within these influences. The very fact that it is orbiting around the Earth, so near to it and still within its gravitational grip, shows its subservience. Frank A. Brown, of North Western University, once wrote: 'What proof do we have that terrestrial organisms removed from the rhythmic geophysical field would continue to maintain their ordinary circadian frequencies? The answer is none. It will continue to be none, until man has either devised chambers which negate all geophysical influences – an unlikely thought, bearing gravity and radiation in mind – or he has placed space vehicles sufficiently far from Earth with a consignment of suitably rhythmic organisms.'

Two main theories about biological clock mechanisms have existed for some time, and both have their admirers. Basically, the pendulum theory suggests that an organism is an independent oscillator, with its own intrinsic timing equalling (roughly) one day. The second idea, called the relaxation oscillation theory, suggests that the organism possesses no such timing but acquires a rhythmic timing from all, or some, of the rhythmic and cyclic geophysical events going on all the time. An ordinary mechanical clock is an example of the first idea. A certain timing has been built into it and that, more or less, is the speed at which it will tick. The human heart is a near example of the second idea. Its speed can be adjusted by its environment, quickened by some hormones, slowed down by others, accelerated by bodily activity, and relaxed by rest. In fact it is only a near example, because the heart has a basic rhythm – akin to the first theory – but this can be extensively adjusted by its environment, as in the second theory.

It is the second theory which has most supporters. This is so much so that Erwin Bünning, who has written (originally in 1958) a classic book on the subject, is extremely offhand in *The Physiological Clock* about the pendulum theory. Arguments in its favour are presented, but half-heartedly. Instead he is a devoted proponent of relaxation oscillation as against harmonic oscillation. Perhaps he is right. Or perhaps, as can be the case when two contrary views have been held for a long time, a bit of each is the final correct solution.

To quote Brown again: 'It is possible that both of the postulated means for the timing of the biological periodisms may turn out to participate jointly in the ultimate explanation of the heterogeneous assemblage of observed rhythmic properties.' That heart analogy may actually be hitting the bulls-eye it was intended to illustrate rather than being slightly wide of the mark. The heart has a basic rhythm, but outside events can stop it, start it, or modify it. This basic rhythm varies between organisms, and the outside events vary in their importance, according to the organism. No one suggests that the ultimate explanation is just round the corner, but this is a new subject and new facts are pouring in all the time. It would be nice to know what makes things tick, how the birds find their islands, and how the bees set off in time for their flowers. The subject is being opened up so fast that, maybe, an answer will soon come. Mankind's estimate of time and how long a job will take can be entirely wrong, after all, despite a plethora of clocks, internal and external.

17 Adaptive Colour

The great naturalists of the last century were both intrigued and delighted by cryptic colouration. They frequently sketched and painted almost indiscernible animals which were crouching among vegetation of virtually identical hue and substance. The apotheosis of all this is a painting by Abbott H. Thayer of a peacock in the woods. The bird's blue neck entirely disappears within the blue of a convenient patch of sky. Its head is equally invisible, matched by conveniently tinted foliage. The great bulk of its tail, with those special peacock roundels, is just detectable amid the variegated leaf pattern surrounding it. With the painter where he is, and the bird so conveniently poised, it is hardly possible to see the creature, despite the fact that the caption makes it plain that a peacock must be somewhere in the painting. The implication is that anyone walking in a jungle, without such precise instructions, would be totally unaware and would fail to see it.

Unfortunately many of the Victorian naturalists were so busy obscuring the subjects of their paintings that they were blind to the many other purposes of colouration, besides that of concealment. Quite apart from the peacock, and other such bright animals, the artists were keen on depicting arctic animals standing invisibly on arctic snow. Such a painting would be side by side with the same animals in their darker summer plumage, now equally conveniently matched with the darker countryside left behind by the snows. Again the implication is that cryptic colouration is paramount, while the animal's ability to remain invisible changes as the snow goes and the snow comes. Such a pair of pictures would demonstrate ptarmigan, mountain hare, arctic fox, stoat and various other species living invisibly and in harmony with each other. What these pictures do not tend to show are the summer coats of, for example, the polar bear, snowy owl, Greenland falcon, and American polar hare. These are all white in winter, but, as an inconvenience for the artists, they are also white in summer.

Likewise, there are arctic and sub-arctic animals which exist in the same area, and which are reasonably matched with their environment in summer, but are certainly nothing of the kind in winter. Examples are the moose, musk-ox, glutton, reindeer and raven. Not one of these animals turns white, however snowy the landscape they live in. To confound the artist further, there are the two main lemming species which fall into both camps. The Hudson Bay lemming *(Lemmus sibericus* or *Cuniculus torquatus)* turns white in winter, while the far more famous – to Europeans at least – Norwegian lemming *(Myodis lemmus* or *lemmus lemmus)* stays the same colour all the year round. The arctic fox, already mentioned as a white species in winter, throws in its own special brand of confusion by being of two kinds, or phases. The sort favoured by the artists is indeed brown in summer and white in winter, but the other is grey or black in summer and often almost black in winter. It can be seen with ease in real life from hundreds of yards away when its dark shape darts over the white land.

The arctic fox highlights the problem. Is whiteness in winter an adaptation? If so, why are only some foxes white foxes and others black? On the other hand, if that white colour in winter is not some form of adaptation, how did it evolve in the first place? Surely, one argues, there must be merits in being either black or white in winter, and one is happy to argue for the Victorian naturalists that white is the obvious colour to be against the snow. Nevertheless black foxes survive, and apparently just as well. Is colour quite irrelevant to them? Perhaps so, because the arctic fox lives during wintertime either on the food left behind by bears or upon its own stores of food buried in the ice, much as a dog has buried its bones in the garden. What need is there for concealment with such an existence?

Obviously there is far more to colouration than meets the eye. A lust for simplicity causes us to think that an animal is content to keep in matching tune with the seasons, and no more. Plainly, such colour adaptation does exist. Plainly it can have considerable advantages in certain cases but, equally plainly, it is not the entire answer. Charles Elton says: 'It is rather interesting to find how emphatically nearly all naturalists who have had wide experience of wild mammals reject the idea of colour adaptation in these animals.' A. R. Dugmore has written that: 'The whole theory of protective colouration in the larger animals may be open to argument, but from my own observations in the field I am firmly convinced that, practically speaking, there is no such thing.' President Teddy Roosevelt, who hunted in Africa and explored in Brazil, made much the same point rather earlier. So did A. Chapman, a few years later, following his experiences in the Sudan.

This introduction to colour is not an attempt to dismiss the importance of adaptive colouration. A bark-coloured moth lying upon the bark of a tree is emphatically benefiting from its camouflage, and scientists have proved the point, but the digressions of this introduction are an attempt to put cryptic concealment in its place, and then to show that there are other possible causes for a particular colouration.

For example, colour is important in temperature regulation, notably for the reptiles, and as every human knows who has the choice of wearing either a white suit or a black one in the tropics. Colour is also important in protecting a body from harmful solar radiation, and human beings become blacker the nearer their traditional home is to the tropics. Colour is important in courtship, and many primates prove this point with the efflorescence of their sexual skins. Colour is involved in the problems of territory: the male anolis, an iguana, has a distinctive colouration when successfully defending his territory. He does not blush with shame when he is beaten; his colour merely changes to the lesser hue of a female.

These examples are not intended to fragment the subject, or to instil complication in place of simplicity, but to show that the colour of an organism involves many factors. (Of course there is always the further possibility that colour involves no factor at all, in that it is partnered genetically by some important characteristic to which it is incidental. The important characteristic is selected for and proved advantageous in the normal processes of evolution while its genetic partner, the colour of the animal, is merely an inevitable accompaniment.) If cryptic colouration held the dominant role that so many of those water-colour artists believed it to hold, one would expect that very many animals would change their colours in tune with the seasons. In fact, due presumably to all the other reasons for colouration, very few do so.

Nevertheless these few are important. They all demonstrate yet another form of adaptation to those fixed astronomical events caused by the sun and the moon, which have been the main subject of this book. There are those examples, already seen, of a response to winter where the animals turn white when the snow comes. There are others which change as the environment changes between the wet and the dry seasons from a greenness to a brownness. Thirdly, there are some animals, active by night and having to lie up by day, which are past masters at concealment. And finally, there are various others, whose lives are totally bound up by the activity of the moon and by its effect upon the tides, that change their colour in tune with the lunar phases.

Firstly, therefore, that change to whiteness. The winter coats of various arctic and sub-arctic species, despite the exceptions, are in certain circumstances an undoubted adaptation to the even whiteness of the winter landscape. The exceptions exist but a general rule still stands that many animals which, because they are either potential predators or possible prey, would gain from being white during the harshness of winter are white at that time. The exceptions are those animals which live on carrion left behind by others, or those which are herbivores and rely for escape upon running away from danger. The raven, the sable fox and the pine marten go for carrion; the musk-ox, the moose and the reindeer go for such vegetation as they can find during this time. These last three, even if they do not run away, all have the capability to defend themselves by standing and fighting it out.

The mountain hare *(Lepus timidus)* of northern Europe is a good example of a creature which becomes white if its environment does so. Its behaviour varies widely; in Scandinavia it habitually becomes white, in the Highlands of Scotland it usually does so, but in the relative warmth of Ireland it remains grey throughout the winter. Similarly the stoat *(Mustela erminea)* only changes in the northern latitudes. Anyone wanting to collect ermine must go to the north of Scotland to be sure of finding it, to northern England to have a chance of doing so, and to the south of England to be almost certain of failure. The weasel *(Mustela nivalis)*, never so satisfactory to an ermine collector because of its relatively insignificant tail, also goes white in northern Europe, but scarcely ever in the British Isles.

It would be nice to be able to state that cold weather or the very sight of snow causes all these animals to go white, but this is far from being the rule. Arctic foxes brought down to the relative warmth of the London zoo still whiten every winter, and stoats often become white in the autumn long before any snow has come and when such a conspicuous mistake of timing is very obvious. It used to be thought that cold was the actual cause of their whitened fur, but this is so only very rarely, as with certain Himalayan rabbits. The current presumption is that the whitening is an adaptive feature. It is the result of a mutation, it oscillates with the seasons, and it need have no further cause. It is a striking adaptation, but then the falling of snow and the persistence of that snow is an outstandingly striking event.

The fact that the countryside suddenly changes from brown earth, from lichen-covered rocks, from the semi-green of a northern vegetation to the universal whiteness of a snowscape is certainly the most dramatic seasonal change on Earth. It is also prolonged. Neither the temperate world nor the tropics can demonstrate such a change, even though leaves come and leaves go, even though the rain falls and then dries up. The sudden change when a desert blooms, when the dry sand becomes a carpet of flowers, is as dramatic as that snowy mantle but far less prolonged and far less universal.

These lesser temperate and tropical changes are partnered by far less animal colour change than in the Arctic. The difference between a European beech wood in summer, with its soft dappled shade, and in winter with its forest of bare poles, is striking, but what animals change colour in response to it? Scarcely any. The insects and other arthropods, for whom colour concealment is vital, have all gone with the leaves. The survivors, hidden, dormant, torpid, or even active, may change their behaviour totally, but not their coat – save for one or two exceptions.

There is, for example, the fallow deer *(Dama dama)*. In the summer all these deer wear their dappled spots, a distinctive uniform often worn by the young of various other mammals. Adult fallow deer are undoubtedly made less conspicuous in summertime, when their own dappling blends with the uneven lighting of the forest floor, but such spottiness would be strident in winter, almost as strident as the zebra's stripes. (So many zebra pictures cause the caption writer to eulogize over the extraordinary camouflage inherent in the animal's markings that one wonders if the myth will ever die. Out on the plains where the zebra lives the animal is outstandingly conspicuous – at least to human eyes – and would be far less noticeable if it were the dun colour of most of the plains game in that environment.)

When winter comes the fallow deer lose their spots, and they become a dull and basic brown. The Japanese deer *(Sika nippon)* also live in forests which are stripped annually of their leaves, and they too lose their spots when the leaves go. Dr Hugh Cott, in his monumental work *Adaptive Colouration in Animals,* makes the extra point that the production of white fur is not a necessary accompaniment to one particular season. Within the Arctic any change to whiteness happens before the winter, but in the deer forest any whiteness to be lost goes with summer. This kind of point makes it all the harder to give these changes simple explanations, such as the arrival of cold weather.

Insects in the polar and temperate worlds tend to disappear from the face of the Earth in winter, and hide in its crevices. Or pupate. Or exist as eggs. The need for concealment does not arise in the presence of the greater hazard of a cold winter. However, in the tropics the insects are largely active throughout the year, both in the wet and in the dry times. Various species respond to this annual change by having two forms, one for the rainy season and another, usually better at concealment, for the dry period. Sir Edward Poulton drew attention to this point in his classic book *The Colours of Animals,* published towards the end of the last century. He was particularly struck by certain butterflies (such as *Precis seasmus* and *Precis antilope)* whose underwings are brightly coloured and conspicuous in the wet season but dull and cryptic for the dry months of the year.

The possession of a bright underwing is common in many insect groups, such as large numbers of grasshoppers, as well as the butterflies and moths. They fly away with a flash of colour that disappears the moment they settle. It is strange that this particular feature should thus vanish for half the year. Poulton concluded that the dry season is an occasion when the onus is on concealment. With fewer insects about, with keener predators in consequence, with torpid states more the rule, and with each insect living longer before it is replaced by the next brood, the priorities are on subsistence, on silent and cryptic behaviour.

Therefore the flash of colour, the diversions and the alarm calls have to go. Conspicuousness may form a good alliance with distastefulness as food in the wet season, when there is food to spare for the insectivores and when their predators can pick and choose; but the combination is less satisfactory in a time of scarcity. It is then better for the prey – even the distasteful prey – to lie low, and not to risk attracting the attention of a predator, however much the very same predator may have been warned off by flashy colouration a few months before.

In temperate Europe there is a similar lack of aposematic insects (those with warning colours) in winter. Wasps, ladybirds and others, splendidly conspicuous in summertime, hide their bright lights under any available bushel in winter. With insects – according to Poulton – scarcity of food causes all the normal rules to be broken. Despite the colossal tropical change from the abundant rains to the parched aridity of the dry season, and from the warm months to the cold, there are few examples of tropical colour change from the good times to the bad. Poulton's argument is convincing – for those that do change colour – but the fact that so few of them do so is even more convincing in helping to dismiss the importance of cryptic colouration for most animals.

However there is a mixed group of creatures for whom this generalization does not apply. It consists of all those whose nights are active but whose days are spent in a fairly static manner making concealment extremely important. This is not a response to the annual seasons but a consequence of that other equally remorseless astronomical event – the Earth spinning on its axis. Night turning into day is too quick for most animals to transform their colour (although the section soon to follow on the crab *Uca* shows that rapid responses are not impossible) but it is relevant to colouration whether an animal is nocturnal or crepuscular, active at night or at twilight.

Think first of all of the nocturnal or evening birds, the owls, night-jars and the woodcock, and then remember their colouration. The woodcock's daytime behaviour, coupled with its dappled grey colour, makes the bird almost impossible to see until, with a sudden flurry, it hurtles off through the trees. The night-jars, which fly about at night looking more like fruit bats than birds, are pastmasters at camouflage. The woodcocks hide in the undergrowth but the night-jars often hide – if that is the right word – out in the open, sitting very tight, and being outstandingly unobtrusive, even if they are known to be in the area. Prof. P. A. Buxton, in his *Animal Life in Deserts,* tells of night-jars near Baghdad which hid themselves – it must be the right word – on bare and open ground. A particular species, *Caprimulgus aegyptius,* was always located in a certain area, but never to be found. It could be disturbed by walking over the area, but it could never be seen by this most distinguished and observant scientist before it actually flew into the air.

Owls are of all shades, white in the north, sandy in the deserts and tawny in the woodlands. Their colour is in keeping with their naturally unobtrusive behaviour. The screech owl of the forests of North America, when confronted by danger, makes slits of its otherwise conspicuous eyes, draws in its feathers to reduce its apparent bulk and form, and then stands more like a bittern in the reeds than any owl. The eagle owl, a long-lived desert bird, has the self-preserving habit of lying flat. And of course, there are owls like the barn owl which retreat during the day into any cavity which can take them. The kiwi, another nocturnal bird but without an owl's talons or a cryptic colouration or even the capability of flight, is a good example of evasion; indeed evasion is its only recourse. It spends all day in a burrow, and has therefore managed not to become extinct despite its apparent unsuitability as a candidate for survival.

In short, if nocturnal activity is partnered by diurnal inactivity, either cryptic colouration or total burrowing from predatory eyes are the two principal alternatives. Of these two the number of animals that sit it out on the surface, such as the night-jars and the woodcocks, are in the minority. The number with holes to go to, bearing in mind the great quantities of nocturnal rodents that retreat from sight, quite apart from quirks like the kiwi, far outweigh those that rely upon skilful camouflage. So, once again, despite the tremendous change between night and day, and between a bright world and a dark one, any colourful adaptation to this great change is slight. The dominant recourse is to evade the problem and live in a place of such concealment that colour is as unimportant by day as it is by night.

The only other time when birds have to sit still is during incubation. A nocturnal bird needs concealment by day because of its inactivity. An incubating bird is similarly inactive, and in similar need of concealment. This point is often emphasized by the colourfulness of the bird partner not doing the incubating, for the females are usually a duller colour than the males. As a general rule, whenever a male is gaudy, he does not help in sitting on the eggs. Of course the nest sites of many birds are not particularly vulnerable during incubation so that concealment is not important. Those which nest in holes or which live in conspicuous communities do not set store by camouflage; it is the small bird on its own in an open nest which cannot afford to be conspicuous. An apparent exception is the painted snipe, because the female is gaudy, with bright green and orange added to its brown, while the male is dull. In fact, this exception bears out the gaudy-male generalization because *all* the incubation is done by the male; the female does none. The female ostrich, a dull-coloured bird, certainly does all the daytime egg sitting while the black and white male is free to run about. However, there is a theory that the male takes over incubation at night when its more strident colouration provides a better camouflage in the moonlight than the female's various shades of grey.

At all events there is an association between inactivity and inconspicuousness. Nevertheless, among the mammals and birds there is no example of a creature which can change its colour with the changing days, and there are relatively few which change with the seasons. Environments change with the seasons and with the time of day, but warm-blooded animals do not, in general, follow suit.

It is different lower down the animal tree. There are many reptiles and amphibians which rely upon escaping detection, rather than possessing any form of defence once detected, and some of these animals can change their colour. When this happens they are reacting in response to changes in their environment other than those caused by the seasons and the days. There is a distinction here. A chameleon, for example, exists only in the present and changes its colour according to the existing situation. It could never react like those premature stoats who acquire their white winter coats by a whole body moult long before any snow has fallen. A chameleon's uniform may be browner during the bulk of the dry season than its bright green colour of the previous rainy period, but this is not a built-in cycle. It is an *ad hoc* response to the present.

It is only among the invertebrates that colour change can be both responsive and innately in tune with the seasons. The fiddler crabs, such as *Uca pugilator* and *Uca minex,* are exciting creatures because their behaviour is set by the moon and yet their colour change is in time with the sun. It is a remarkable state of affairs whereby these arthropods have adapted themselves both to the lunar cycle and to the solar cycle. Mankind has traditionally wanted to fuse the two, to have a fixed number of days between each full moon, but this cannot be achieved any more than a fixed number of full moons can be fitted into each full year, that other early human aspiration. The fiddler crabs, so called because of their single gigantic claw's slight resemblance to the instrument and the way they wave it about, have not merged the unmergeable. Their instinctive response to the moon has remained distinct from their instinctive response to the daily cycle of the Earth's revolutions.

This means that the crab emerges from its burrow at an appropriate time for the state of the tides and with an appropriate colouration for the state of the day. It is an excellent arrangement. The emergence, caused by the moon, occurs a certain time after the water has receded from the area where the crabs spend the higher part of each tide. They spend these high-tide hours in burrows beneath sand then covered by water. It is no good emerging too soon, for the wet sand would then be insufficiently firm to support their escape holes. They have to wait for the sand to harden, so that the holes have a chance of remaining good until such time as the crabs will want to scuttle down them again in the face of the advancing water. Because high tide always occurs at a different time each day, all this crab activity, the emergence and then the retreat, has to occur entirely independently of the sun.

Nevertheless the crab's chromatophores, or colour spots, either contract or expand so that the animal is in a suitable livery for the time of day when it does emerge. The rhythm of colour change continues in obeisance to the sun, and the crab's activities are entirely dictated by the lunar rhythms and by the consequent state of the tides. The solar rhythm of changing chromatophores is so engrained that it will persist for thirty days in experimental conditions entirely lacking in outside influences.

Once again a biological clock can be demonstrated and, once again, by dropping the temperature. Should the crabs be kept at freezing point for, say, six hours, and then warmed to normal temperatures, they will soon start to behave normally once more. However, as with Hoffman's starlings, they will then be six hours out of phase with the sun. The irregularity of the laboratory conditions will have shown up the persistently regular endogenous pattern. Without such crude laboratory interference, the fiddler crab is entirely capable both of being active in accordance with the moon and of being colour adaptive in accordance with the sun.

Now to the mechanisms behind the simplicity of the statement. By no means are they clear cut. Not only are there several kinds of chromatophores, but it would seem as if each type of colour spot is controlled by a distinct hormone. It is also possible that the hormonal control is sometimes not so much due to the presence of a particular hormone as to the counteracting presence of two hormones, in which case the chromatophore will respond most to the hormone which happens to be dominant at the time. One of such a pair of antagonistic controls will be enforcing pigment dispersal while the other will be concentrating the pigments.

Where are the hormones produced and what causes them to be secreted? In the crustaceans there are various areas where hormones are produced, and there are three which appear to be more important than the others. Firstly, there is a lump of tissue within the eyestalk called the sinus gland (because it exists on the edge of a blood sinus). Secondly, there are the post-commissure organs, equally modest lumps which could readily be mistaken for part of the nervous system because they are directly attached to it as small swellings. Thirdly, there is the pericardial organ, reasonably named because it consists of a small bundle of tissue enclosing a lot of nerve endings which lie on top of the crustacean heart. These three are certainly not the complete list, for various other similarly small blobs of tissue have been indicated experimentally as hormone secretors in many other invertebrates, but these three are generally important, and the first two of them are considered to be of principal importance in, for example, the prawn *Leander*.

As can be imagined, the process of crushing up a small lump of tissue from an invertebrate, trying to extract the potent secretion of that lump of tissue, injecting that secretion, and noticing any colour change which results can be a finicky business, and not always successful. Another approach has been to excise the organ in question and then to observe the effects of such surgery. Once again, such a procedure on such a type of animal is not always commensurate with simple and effective conclusions.

Quite apart from all these problems there is the major additional inconvenience that many invertebrates, such as *Leander* and the crab *Uca*, change their colour for all sorts of reasons. There are the fundamental colour rhythms already mentioned, which arise in response to the sun (as with *Uca*) or to the moon, but there are three other important kinds of response.

There is the effect of illumination itself. In usual conditions bright light will tend to disperse pigment. This can be proved by shielding a small portion of a fiddler crab's leg, and then shining a bright light on to the whole animal. The portion shielded from this illumination will remain darker. Secondly, there is a response to temperature. In general, a higher temperature means a concentration of red pigments and a dispersal of white pigments. The effect of this rearrangement is to cause the animal to absorb less radiant energy, and therefore to alter and control its temperature. Thirdly, there is what has been called the albedo response. The chameleon is probably the most famous animal in the world for this type of colour adaptation, but many invertebrates also practise it. In general its effect is to match the animal with its environment. If the creature is on a light background it will respond by becoming lighter, and even if only half the animal is on a black background that half may become darker.

These three other responses, associated with brightness, temperature and camouflage, can completely swamp the effects of the ordinary diurnal rhythm. *Uca*, for example, may emerge from its sandy burrow appropriately tinted for the time of day and for the height of the sun, but the heat from that sun, or the illumination from it, or the brightness of the sand, or all three, may then change that appropriate tint into something quite different. However, this different colouration may be even more pertinent and appropriate when all factors important to the animal are borne in mind.

Therefore, so far as annual and daily rhythms are concerned, the invertebrate animal is likely to be at a consistently different colour according to the time of day and according to the time of year, but for all sorts of reasons. The sun is less bright in winter, the radiation is less warm in winter, the incident and reflected light is a whole lot less in winter, and the background may be quite a different colour at that time. Such rhythms as do exist may be secondary to all these other major considerations, and therefore hard to detect. It is only when the animals have been taken away from the many-sided and conflicting effects of their normal habitats that the rhythms, such as they are, can be demonstrated. Nevertheless, this is not to belittle them. Undoubtedly there are rhythmic colour changes in the invertebrate world, however hard it may be to elicit them and however much they may be trampled upon by other effectors.

The vertebrate world has similarities but also major differences. A prominent similarity is that being vertebrate, or even being warm-blooded, does not affect the various reasons for colouration. Therefore it is as hard as with the invertebrates to sort out rhythmic colour change from the more immediately adaptive kind. Another likeness is the vertebrate possession of three specialized types of pigment cells, called melanophores, erythrophores and xanthophores (specializing in black, red and yellow respectively) very like the invertebrate chromatophores. As before, a colour can be made more strident by dispersal rather than concentration, and within each particular cell is a ramifying network along which the pigment can be dispersed and from which it can be concentrated. It is in the control mechanisms that the principal differences exist. As before, a colouration may be controlled directly by nerves, or by nerves and hormones, or by genes. As before, it can be difficult unravelling the three, but the hormones themselves arise in a different manner in the vertebrates. This might be expected, bearing in mind the extremely dissimilar nervous and glandular arrangements in these two great divisions of the animal kingdom.

Melanophores, often called melanocytes, are the cells which manufacture melanin, the black pigment. Probably these cells are identical – in so far as that word is applicable – to the chromatophores of the invertebrates, but in the vertebrates they are controlled by a secretion from part of the pituitary gland. The anterior pituitary has already been referred to (notably for its influence upon the sexual cycle) and here a part of the same gland is exercising similar control. Although extremely small the pituitary gland is made up of more than one part. While the anterior pituitary is so crucial to the sexual cycle, it is a portion just behind it, called the pars intermedia, which secretes the melanocyte-stimulating hormone. As with the sex hormones the existing lack of knowledge about their chemistry has led to functional simplicity in their names; MSH is the name given to the pars intermedia secretion which acts on melanocytes. Other varieties of MSH have been discovered, and they too have been given initials, but it is not yet known whether any hormone acts antagonistically with MSH, or whether melanin dispersal is controlled simply by the presence or absence of this single hormone. Once again there is the experimental difficulty that the experiment itself, the act of injection, can cause colour change quite independently of any substance which may be injected at the same time. (In brief, stick a pin into someone and he will go red, whether any chemical went in with the pin or not.)

Anyhow, MSH is a definite hormone in cold-blooded vertebrates, for its presence leads to pigment dispersal (i.e. darkening of the skin) and its absence to concentration. Many fish, amphibia and reptiles can be made permanently light-coloured by cutting out their pars intermedia. They do not remain either permanently red or yellow in consequence because MSH, the misleadingly named melanocyte-stimulating hormone, also acts on the other colour cells. Perhaps MSH is more than one substance, and each causes more specific reactions than is appreciated at present. Perhaps . . . ? The science of hormonal control has a long way to go.

MSH seems to behave differently with warm-blooded animals, being more of a stimulator of melanin production than of its ebbing and flowing from a particular spot. This makes colour change a rather more permanent affair, for the melanin will remain until some equally permanent effect can cause it to disappear. This may explain why diurnal and short-term rhythms seem to be commoner in cold-blooded animals, and why the regular colour change that does occur in warm-blooded animals is of a long-term nature, such as the seasonal alteration from black and white in those arctic mammals and birds.

Unfortunately, although a lot more is now known about skin darkening and lightening, the higher an animal exists in the evolutionary tree the less is known about its colour change. Little more is known today about the mechanisms lying behind a fox's seasonal whiteness and blackness than was known by those Victorian artists who thought primarily of concealment. Most facts are still equally well concealed. Sex glands and hormones are thought to complicate the problem; so, too, the thyroid gland. As the hypothalamus has such an effect upon the pituitary gland, and as the pituitary is so influential upon so much of the endocrine system, and as hormones from this system interact in a hideously complex action, it is easy to take refuge in a statement that the hypothalamus is probably at the back of seasonal change.

Somehow or other the hypothalamus notes the passing months despite climatic unevenness and irregularity, and thenceforth causes colour changes which may or may not necessarily be advantageous to the animal. Therefore, somewhere, as with so many other aspects of this subject, there is a clock at work. There must also be a system keeping that clock accurate, and there is tissue waiting to react to nervous stimulation from it, to go dark and white, to respond, to change.

The story of seasonal colour change is essentially the same tale as with hibernation, migration, or sexual activity. It is yet another response of the animal to its rhythmic environment, and the animal makes adjustment via its sense organs, its clock, its control devices, and its behaviour. Unlike the other three, it so happens that colour change in response to the seasons is of marginal importance. At least, this appears to be so when set beside all the other factors impinging upon the desirability or otherwise of a particular colouration. The ability to be of a certain colour, it seems, is too important to be left merely to the static requirements of camouflage.

Mankind, and his varied hues, provide an effective tailpiece to this story. Admittedly Eskimos are white in a white world, and Africans are black in the dark forest, but no one can pretend for a minute that concealment is behind this colour change which parallels the lines of latitude. Instead, *Homo sapiens* is a species in tune with a generalization which says: the nearer the tropics, the darker the creature. But why is the Negro black when a white suit is a good tropical outfit?

Without doubt a black man sitting in the sun absorbs more heat and becomes hotter than a white man. Similarly a black man will lose more heat during a cold night than a white man. Therefore, in terms only of adjustment to heat and regulation of temperature, a black man's body has to work harder than a white man's when both are in the tropics. Man is thought to have arisen in the tropics, and the white man's relative lack of melanin dispersal is thought to be a relatively late development. So there has been plenty of evolutionary time for the black man to lose his blackness suggesting that there must be a reason for retaining it.

So far the most attractive theory involves protection of the inner layers. After all, man is the naked ape, and his skin does receive the full ground-level radiation from the sun. A black skin has little transparency, and absorbs a lot of radiation close to its surface. A light skin is more transparent and, although it absorbs less, it absorbs more at its lower levels. As vulnerability and sensitivity increase with depth, and as the outer layers of skin are already dying or dead, the black skin may have its blackness as a shield. In other words temperature is something which can be regulated more effectively, and put right more readily, than solar radiation damage. No one can prove this point, for it involves experimental work on human beings and the deliberate infliction of damage beneath the skin; but it is a nice theory, none the less. However it is also a digression because, so far as is known in a subject steeped in unknowns, there is no innate daily or seasonal variation in any one person's colouration, however colourful he or she may be.

18 The Plant World

It may appear contrary to encounter the seasonal situation of the plant world only after considerable discussion mainly on the various responses of animals to the changing scene, and indeed it is contrary, but there are reasons. Much more work has been done on animals, and the animals have often indicated to botanists how their problems might be unravelled. Also animals are mobile, and can respond more excitingly – as in migration – to their seasonal planet. A static and dormant tree is far less arresting in appearance than a lively little animal's sudden transformation into a metabolic lethargy. A seed's quiet acceptance of the wintry situation, and then its arousal into germination is, to many observers, less intriguing than the springtime rituals of the animal world.

Nevertheless, despite the obvious dissimilarities between plants and animals, there are some fascinating cross-references. Most cold-temperate trees shed their leaves in winter, and then carry out their growth in the summer; but the hot-temperate world witnesses a reversal. Photosynthesis in the Mediterranean zone, for instance, is a wintry business, and so is growth; summer is a time of dormancy. So too with animals. Cool-temperate animals realize most of their activity and reproduction during the warmth of summer, and merely subsist throughout the winter. In the hotter climates both the cooler nights and the winters can be preferable for growth and reproductive activity, with the heat of the day being a time for torpidity, for aestivation.

Similarly the temperate world is not just an intermediate place between the tropics and the poles. Hibernation only occurs in the temperate zone (save for the northern ground squirrel), and so does annual leaf fall; neither the conifers nor most tropical trees lose their leaves every year. Yet another parallel is the difficulty of the British climate for both vegetation and animals. As a consequence the British Isles are poorly represented in numbers of species when compared with places either farther south, where the climate is warmer, or farther east on the same latitude, where the winters may be more severe but are customarily more reliable. The ability of the British Isles to have hot days in January and snow in April can be exceptionally damaging to both kinds of community.

A final parallel is that sexual organisms, whether plant or animal, have to synchronize the activity of both male and female partners, and the mechanisms of doing so are extremely similar. The ferret takes unconscious note of the passing days, and its hormones act as messengers, transforming the fact of changing daylength into the fact of sexual synchronization. So too with plants. They operate in accordance with the calendar. Seasonal variables are detected. Hormones are manufactured. The reproduction areas are transformed. Sexual synchronization is achieved, and growth and budding and flowering and seed-making all take place in accordance with the seasons. The oak tree's coming into leaf at the same sort of time each year is no more magical than is the swallow's regularity of arrival or the starling's nocturnal incubation.

Amazement was expressed in an earlier chapter that no one had realized, or proved, the importance of changing daylength in the animal world until the period between the two world wars. It was considered astonishing because changing daylength is the only reliable factor throughout the year, the only one which can prevent a January day being mistaken for some later date. The equinoctial gales can apparently blow at any time, rain can be unseasonal, and warmth can be quite irregular, but daylength is as accurate as the calendar because it is the calendar. It is possibly even more amazing that botanists did not associate daylength with plant processes from the very beginning. Daylight is important to an animal of course, but it is crucial to a plant. An animal can get by without much daylight, or can even become nocturnal, but a plant cannot undergo photosynthesis and make use of solar radiation except in the presence of sunlight. However, less reasonably, no one put two and two together until very recently that lengthening (or shortening) days are not only crucial to plant activity but act as the trigger for various aspects of that activity.

It was W. W. Garner and H. A. Allard of the US Department of Agriculture who first linked daylength with flowering. The year was 1920. (Thus there was only a narrow margin of four years before William Rowan did the same kind of correlating work with the animal migration of the junco.) Their daylength discovery is now called the phenomenon of photoperiodism. Others ought to have made the discovery, in that they were getting similar results, but these others tended to interpret their findings on a nutritional basis, believing the light to be changing sugar-balance, etc., instead of acting as a trigger for hormone production. Garner and Allard were working with Maryland Mammoth tobacco plants, and were intrigued that these would only flower in the short days of winter whatever the date of their sowing. Later the scientists covered up some of these tobacco plants for part of each summer day, and this act led to summer flowering. They then prevented winter flowering by artificially increasing the winter's short daylight hours.

These two men had the insight to perceive the actual role of daylight, and they repeated the experiment with many other flowers, before concluding that there are three types of plants. Firstly, there are 'short day plants', like the Maryland Mammoth tobacco, that are stopped from flowering by long days and short nights. Secondly, and conversely, are the 'long day plants' that are stopped from flowering by short days and long nights. Thirdly, and contrary to both these groups, are the 'daylength neutral plants' that flower whatever the length of day.

In the first two cases, the longness or shortness of a particular daylength is critical to the plant. For example, the cocklebur *(Xanthium pennsylvanicum)* is a short day plant that will never flower if days are kept permanently with sixteen hours of light followed by eight hours of darkness. However, if this plant is given just one dose of a day when darkness lasts for more than nine hours, and whole daylight therefore lasts for less than fifteen hours, *Xanthium* will leap into floral activity, and will continue to flower for at least a year, whatever the daylength situation that follows the single dose. An example of a long day plant is *Hyoscyamus.* Its critical daylength is ten to eleven hours, and it will flower whenever daylength is longer than this time.

It was soon realized that, although flowering follows differentiation at the apex of the plant, it is not the apex which needs to be subjected to the daylength; it is the leaves. Sometimes the stimulus needs to be given to only one leaf, or to part of one leaf, for the full process of efflorescence to occur. (There is again a parallel with the ferret of Chapter 3. It is not the animal's gonads that need to receive the daylight but its eyes.) Once the stimulus has been given, and the developmental process has been initiated, this will continue without further stimulus. With some plants it is even possible to subject a single, detached leaf to the appropriate daylength, and to cause flowering in another plant merely by grafting that one leaf on to the other plant. The grafting need not even be between the same species to be successful, and the two plants need not even be in the same daylength category. The long day plant *Hyoscyamus niger* can be grafted on to the short day Maryland Mammoth, and thereby induce it to flower in long days. Grafts by short day plants on to long day plants can be made to have the reverse effect.

All this experimentation indicates that a hormone is involved. For the leaf to receive the stimulus, and for the plant apex to act on it, is sufficient indication, but the cross-grafting further suggests that it is the same hormone which is involved. This flower-inducing hormone, called florigen for convenience, is believed to travel from leaf to apex along with a good many other substances like the sugars. However, unlike the sugars, it travels relatively slowly. Sugar movement measured in one plant was around 3 ft an hour, while florigen in the same plant moved

½–1 in an hour. Reasonably enough, florigen does not travel from a graft into the host plant until the graft union has been properly formed.

There used to be an argument, before the notion of a hormone became so firmly implanted, that the leaf/apex association was formed as a kind of by-product of the chemistry of photosynthesis. Somehow or other the nutritional benefits resulting from exposure to sunlight enabled the apex to start growing. It is now known that the amount of light needed to trigger off apical growth is insignificant by comparison with the demands of photosynthesis. One foot-candle is sufficient for many species, either to cause flowering or to prevent it. (A foot-candle is the amount of light exactly one foot away from a standard candle of a particular wax, a candle not greatly dissimilar to the ordinary domestic kind. One foot-candle is about twenty times as bright as bright moonlight and therefore extremely modest radiation by comparison with sunlight itself.) Certain plants have shown an ability to adjust their flowering cycle if daylength is extended by only one-tenth of a foot-candle, and the nutritional benefits arising from such a small ration of light must be minute. However, some sort of photochemical reaction is involved because soy beans will not flower, whatever the light they receive, if they receive it in air devoid of carbon dioxide.

In short, what florigen is, how it is manufactured, how it travels, and how it triggers apical growth are all unknown, but its stimulus is certainly effective. That single dose treatment of *Xanthium*, causing it to flower whatever happens to daylength afterwards, suggests that the word trigger is perfectly apt. Perhaps it causes florigen to be manufactured at the apex with this species, and perhaps not all the species operate in the same fashion because, with some, a lack of continuing stimulus at the leaf causes a subsequent lack of flowering at the apex.

Yet another unknown, a consistent difficulty in all hormone problems, concerns inhibition and activation. Is florigen actively promoting flower production, or is some hormone secreted during the unfavourable days which inhibits flower production? Experiments have been made on different plants which support both ideas. Plants have been given a go-stop-go daylength pattern in which the number of go days equalled the number necessary to promote flowering. With some of them the plants have flowered normally. With others the stop days have apparently over-ruled the go days, thus possibly suggesting the action of an inhibitor.

But this experiment may suggest nothing of the kind, for the role of the intervening dark periods in plant life is certainly not yet understood. Plants, whether short day or long day, usually need the dark periods in between each period of light, and the dark period must itself be preceded by sufficient light for the darkness to be effective. With the short day *Xanthium* the long dark period essential for the promotion of flowering must be *preceded* by about thirty minutes of light.

The stimulus will not happen if the long dark period is *succeeded* by the same amount of light, assuming that only a single dose of treatment is being given. Also, however correctly long the long dark period may be, and however correct the light period before it, the effectiveness of the dark time can be vitiated if it is interrupted by quite a modest amount of light. An interruption of a few minutes of ordinary light, or even a second of very intense light, will destroy the dark period's effectiveness. Therefore, whether inhibiting hormone or promoting hormone is involved, the reversal process, whatever it is, is certainly extraordinarily rapid. Such speed of action is customarily associated more with the zoological than the botanical world.

It is only with the short day plants that the period of darkness is essential. The long day plants need only light, and can dispense with the dark period entirely. Therefore a real distinction exists between these two plant divisions. In fact, a long day plant comes into flower most rapidly in continuous light. Not all long day plants are the same, but with many the effect of long days is not forgotten even if there are short days in between. Annual beet needs eighteen successive long days if it is to flower under normal conditions. Under the abnormal conditions of an experiment it will still flower if ten long days are given, then thirty short days, and finally eight more long days, bringing the total up to the necessary eighteen.

What light do these varied results shine upon the hormone system? As yet the hormones themselves are still unidentified, but there are theories about the possible order of events. F. G. Gregory has suggested a hormonal sequence for the short day plants. Let there be substance A formed in a photochemical fashion, making use of both light and carbon dioxide. A lot of light will convert all the A into X, a substance which can best be forgotten in this argument because it plays no part in flower induction. During darkness A is turned into B, slowly at first, but then more rapidly because the length of darkness is important for these plants. (One must assume that B can be turned back into A in a trice when one remembers the effect of that single flash of light in the darkness. Anyhow, under normal circumstances, there are no flashes in the middle of the night, and so A becomes B.) Subsequently B leaves the leaf and is no longer sensitive to light, but can be turned back into A if the rest of the plant is not producing B in similar fashion. Finally B becomes C, a light-stable compound which actually does the job of inducing the apex to grow and produce flowers.

Maybe this procedure is not the right one. If not, there are probably more steps to the process rather than fewer. At all events the word florigen conveniently describes the flower-producing hormone. It does not attempt to explain what the difference might be between eight and a quarter hours of darkness and eight and a half to nine. *Xanthium* will not flower if only exposed to the former, but will do so if given eight and half hours or more of darkness.

With such precision being involved the postulation of yet another biological clock has been inevitable. F. C. Salisbury (formerly of Kew Gardens) and others have been asked whether this timekeeper is of the hour glass or pendulum kind. An hour glass needs sufficient time for various processes to be completed, while a pendulum suggests that there is an internal rhythm in the plant oscillating between light promotion and light inhibition. How such an endogenous rhythm operates is quite inexplicable as yet, but opinion seems to be in favour of a pendulum rather than an hour glass. Perhaps, as with the animal kingdom, the rhythm will prove to be an innate feature of every cell, a protoplasmic ebbing and flowing rather than a specific device. In such a basic aspect of life it is possible that plants and animals are similar. Therefore a scientific breakthrough on either side ought to illuminate both sides. The animal sciences have tended to call the tune for the plant sciences but, just as Garner and Allard came before Rowan, the true nature of the biological clock may first emanate from the plant world.

In the plant discussion so far temperature has not been mentioned. Although relevant it has been deliberately held back to this point, partly to emphasize the all important factor of daylength. Temperate man, obsessed with the warmth of spring and the departure of wintry days, has always been too prone to believe temperature the prime consideration, with moisture being an equal first or a good second. It has already been suggested that this obsession must have been partly responsible for the delay in discovering the importance of daylength, but even today, half a century after that discovery, daylength is probably running a poor third in most gardening and agricultural minds as a trigger of plant activity. No one can deny the importance of temperature, but the object of this paragraph is to stress that many are still disregarding the importance of daylength.

With temperature there are, once again, parallels between the animal and plant kingdoms due to the fundamental law of nature that heat speeds up chemical reactions. For example, greenhouses certainly do not change daylength (unless there is lighting in them) but their increased warmth – or lack of cold at night – does affect growth. Bulbing in onions not only begins sooner, but proceeds faster during warm days, as against cold days of the same daylength. Warmth at night is here more important than warmth by day.

Coupled with the broad thesis that things happen faster when they are warmer is the complication, already referred to under biological clocks, that no time-piece is good if it is unduly influenced by temperature. There has to be a system of compensation. In other words a rise in temperature will both speed things up and, on other occasions, will not speed things up owing to the system of compensation. Therefore the result of an increase in temperature is certainly not straightforward.

If small onion bulbs (called sets) are primed to start flowering, and are then planted in early spring, the prevailing temperature, particularly at night, will influence their development. They will either bolt, as the flowering process is rapidly carried out, or they will develop much more bulb tissue. Which of these two happens is, of course, enormously important to the onion industry. Temperature and light have to be skilfully manipulated. (Onions are contrary to almost all other plants with a storage organ, because the bulb is formed in response to the trigger of long days, rather than short.)

The Dutch have been outstandingly adept at the business of controlling light and warmth in their cultivation of the famous flowering bulbs, like daffodil, hyacinth and tulip, and in making everything happen in the shortest possible time. The procedures are extremely complicated. Darwin tulips, for instance, are frequently lifted in June and then stored at 68° F (20° C) for three weeks. Then they have three weeks at 46° F (8° C) and the three following weeks at 48° F (9° C) before being planted early in September. In darkness and at 48° F (9° C) they then put out their first leaves. After seven weeks of this temperature they are gradually warmed – three weeks at 55° F (13° C), three weeks at 68° F (20° C), and three weeks in the light at 73° F (23° C).

By that time, less than six months after the lifting, all the flower bulbs will be tinged with the forthcoming colour. This complex cooking of the books by some Dutch bulb growers should certainly stress the importance of temperature, but it should also indicate that the role of warmth is not entirely clear-cut. Trial and error rather than a theoretical understanding have produced this particular recipe for Darwin tulips.

Vernalization is another treatment which has oscillated from simplicity to complexity. It now means the speeding up of flowering by adjustment of the temperature at some earlier stage of development, but the word originated entirely in association with winter cereals. In general these come in two kinds. There are the spring varieties, sown in spring to ripen the same year, and the winter varieties, sown in autumn to ripen the following year. Not only are the winter cereals hardier, in that they can withstand temperatures which would kill the spring varieties, but they need the winter to achieve their development. If sown in spring they still do not produce ears until the following year.

The simplicity of this story has been confounded by various twentieth-century discoveries. Winter cereals will ripen in the same season if kept cold (just above 32° F (0° C)) during their germination for one month before planting, and cold treatment without the water for germination will do just as well. Also winter rye grown throughout the winter in a greenhouse (and therefore warmer) will still ripen the following summer. Even more perplexing is the fact that the influence of daylength depends upon the number of leaves each variety of cereal has to produce before it can produce ears. Winter treatment, whether natural or artificial, reduces this critical number of leaves, but everything is interdependent and the low temperature effect is itself influenced by daylength and vice versa. Here is O. V. S. Heath of Imperial College, London, in the same kind of semantic difficulty: 'Thus shortday treatment applied early has an "after effect" in reducing the leaf number, somewhat similar to the after effect of low temperature but different in magnitude.'

Temperature, therefore, is no one thing, no simple factor, no straightforward guideline to the welfare of a plant. Much depends on when it comes, and for how long, and what it follows. The growth of young tree seedlings shows up this complexity. For them to achieve maximum growth there are differing optimum temperatures for both days and nights. The Douglas fir *(Pseudotsuga menziesii)* grows best if the days are at 73° F (23° C) and the nights are at 45° F (7° C). *Sequoia sempervirens,* the redwood, will grow well even if the nights are warm. *Pinus lambertiana* needs even colder nights than the Douglas, with its growth being optimum if the nights are 39° F (4° C) and the days are 73° F (23° C). *Pinus taeda* needs days 22° F (12° C) warmer than nights, and growth will be considerably reduced if both days and nights are all at one temperature, whether the warmer or the cooler. Eucalyptus, on the other hand, to name quite a different kind of tree, grows best if the days are 75–86° F (24–30° C) and the nights are only 5–9° F (3–5° C) cooler. Presumably the definition of a truly optimum temperature range can be made yet more complex by examining optimum evening and morning temperatures, or by introducing the prevailing daylength as another parameter, or the amount of actual solar radiation, or of moisture. Temperature can no more act on its own than any other requirement.

To temperate man one of the most striking visual effects of the seasonal year is the autumnal loss of leaves, the fall. The resurgence of this canopy in the spring, the full leaf of summer, the loss of this green mantle and the bare poles of winter epitomize the yearly round. Admittedly there is much other botanical change from summer to winter, as annuals die, as seeds are formed to lie dormant, as growth stops; but the seasonal coming and going of the foliage is an outstanding and conspicuous response to changing conditions. The whole tree, the whole deciduous forest, appears to die and then springs so lavishly to life.

It was generally assumed either that cold days caused the leaves to fall, or that there was a lack of nutrients at the same time, and somehow the two of them transformed the scene. It is now known that a photoperiod is involved, that leaves fall in association with shortening daylength. When the days start drawing in it is dominantly this fact, rather than any nippiness in the air or shortage of food in the soil, which is triggering leaf change and then leaf fall. Once again the one really reliable component of the year is the permanent indicator of time's passage, and not the unreliable temperatures or climatic conditions. However, as before, both temperature and climatic conditions can modify the situation.

The reason for leaf fall is assumed to be a method of water conservation during the time of year when photosynthesis is at its lowest. The game, so to speak, is not worth the candle during winter, particularly as the high winds of winter and of early spring can aggravate water loss. Not every tree is deciduous. The conifers in the north have rounded leaves with quite a different transpiration pattern, quite unlike that affecting the maximum surface with minimum bulk arrangement of the typical deciduous leaf form. Tropical trees do shed their leaves, as they become injured or old and as the tree itself both grows and changes shape, but they do not – in general – do so annually. Some tropical trees are emphatically deciduous in that they shed all their foliage from time to time, leaving their branches as bare as an English oak in winter, but they are the exceptions.

Within the cool-temperate world practically every tree species sheds its leaves. There are evergreen exceptions, such as laurel, holly, ivy and the holm oak *(Quercus ilex)*, which really ought to be living down in the warmer Mediterranean area, and the leaves of these species tend to be less vulnerable and less delicate than the traditional deciduous kind. The conifers also do not shed their leaves, with an exception to this rule being the deciduous larch tree. If one travels east from the British Isles into the heart of continental Europe the deciduous trees gradually disappear and are replaced by conifers. The last to survive is the lime, an odd fact bearing in mind its considerable summer expanse of delicate leaves.

Despite the energy involved in growing a new leaf system every year, trees manage to cut their losses to a considerable degree. By the time the leaves drop off, they are entirely dessicated replicas of their former selves. They generally appear dead and discoloured, and it is this dying which creates the superb colours of the autumn scene. Before each leaf death all the proteins and other complicated (therefore valuable) chemicals in the leaves have been reabsorbed into the main body of the tree. A typical leaf loses all its chlorophyll, its anthocyanin, much of its nitrogen, phosphorus, sodium, iron, magnesium and water, and even changes its carbohydrate constituents, before it is discarded by the autumnal process of abscission. A tree behaves somewhat like those garages that cannibalize cars to strip them of all value before consigning them to the scrap merchant. Both trees and garages therefore possess many of the necessary and valuable materials when the time comes for replacement. The scrap merchant mainly receives old iron.

Similarly the ground mainly receives cellulose, and cellulose is easy for the tree to make again.

Abscission is the actual severance of leaf from stalk. Photoperiodism, and the shortening days, have their effect via hormones and these somehow stimulate the resurgence of chemicals into the main body of the tree. The process of cell wall dissolution, which then leads to abscission, is unknown and it is no clearer with other forms of severance, such as the shedding of bark, of fruits, seeds, and flower parts. In virtually every case some kind of injury or senescence precedes the abscission, and the actual shedding of the leaf follows automatically from this damage. The withdrawal of all those chemicals is just another form of damage in that the leaf is undoubtedly inferior once they have left.

With accidentally inflicted damage the amount of injury is important to the likelihood of subsequent abscission. With citrus trees the quantity of leaf fall is directly proportional to the area of each leaf which has been severed. Abscission can also follow if the leaf fails to develop properly. Therefore whatever factors do regulate abscission lie with the organ concerned. The falling of a leaf is voluntary euthanasia rather than suicide, and certainly not murder by the rest of the tree. The leaf itself detects the shortening days and the hormones are liberated within it. They cause the chemical resurgence and, when this is sufficiently advanced, there is cell wall dissolution and the leaf soon falls to the ground.

If a leaf is murdered in the sense that external influences are both strong and sudden, it may not fall. For example, if a cotton leaf is subjected to a moderate frost, it will fall. If the frost is severe, the whole plant will be killed and the leaf will remain attached. Temperature change generally acts in an indirect manner, either via injury to the leaf (as with frost) or via less food. Water is similar. A lessening of the supply can certainly act as an injury, and leaves will subsequently fall, but a total and sudden lack of water will be too blunt an injury, and there will not even be enough water for the abscission process to occur. Cool temperatures are generally associated with leaf fall, but in fact abscission happens fastest at the higher temperatures between 77° F and 86° F (25° C and 30° C). This contradiction occurs because the lack of warmth is both a trigger for an event to take place and, as with chemical reactions in general, a factor that slows down the event. Leaves injured in the high days of summer will be lost far faster than either autumnal leaves or evergreen leaves injured in winter.

Temperature plays a similarly dual role with growth itself. It is axiomatic that there should be warmth, and growth tends to stop at temperatures below 43° F (6° C), but temperature without daylength (and not just daylight) is powerless. Many deciduous trees and southern pines stop growth if the days are less than twelve hours long, and that includes the whole winter period from autumnal equinox to spring equinox. There is also the problem that one daytime from precise sunrise to precise sunset does not necessarily mean the exact length of the day in terms of sunlight, at least as judged by one particular organism in one shady spot. Days differ, and scientific workers in this field tend to blur the issue conveniently by referring to daylength needs both in suitably round figures and to the nearest hour. Also daylight can continue after the sun has set and the day has officially ended.

Experiments with Douglas fir have shown the importance, and complications, of daylength. For example, the fir does not grow if the day is less than twelve hours long, but growth will be virtually continuous if twelve-hour days are artificially maintained and interrupted by single hours of darkness. This arrangement of twelve hours daylight, one hour dark, twelve hours on and one hour off, will produce as much as a standard midsummer day of twenty hours on and four hours off.

Sitka spruce trees are even more demanding in that they do not grow – in experimental conditions at least – if the days are less than fourteen hours long. Norway spruce, yet more demanding still and frequently planted even in the low latitudes of southern England, need sixteen-hour days. As latitude 52° (half a degree north of London) has a sunrise to sunset time of only sixteen hours forty-five minutes on midsummer's day, and as even Edinburgh at 56° N has a time of seventeen hours thirty-seven minutes on the same date, and as Norway spruce grows well in both areas, one can only assume that the tree does not wait until sunrise, or stop short at sunset, in its registering of daylength. After all there is a certain amount of light even after the sun has gone below the horizon and before it rises in the morning.

Resumption of growth is also not entirely straightforward. Catalpa, for example, can be made dormant by short photoperiods, i.e. by short days, but it will quickly start growing again if it is soon given longer days. If the period of short days is longer there will be an interval after it has been put on long days before growth will start. If the short day period is exceptionally long, there will be no growth when given long days again even after an interval. Instead there has to be a cold period, and then a time of long days, before growth can start again. In other words, there has to be a winter – for Catalpa. There does not have to be one for Asian white birch. It will resume growth at any time if and when the days are long enough.

The start of growth in the first place, when dormant seeds spring into life, is not wholly a matter of moisture and warmth. Many seeds are light-sensitive, and can only be triggered when light exists as well. Some can start after a single brief exposure to light; others need forty-eight hours. Still others require a forty-eight-hour light treatment, but will start growing with one hour of light at the beginning of the period, one hour in the middle and one at the end. With seeds requiring both suitable temperature and light to start growth, it would seem as if both systems operate independently. Yet too low a temperature can negate the effects of daylength, however long this may be.

Photosynthesis, the creator of all things – well, nearly all – requires light of course, but also requires different temperatures in different species for this chemical process to be most effective. At 59° F (15° C) seaside bent grass (a cool zone species) can photosynthesize at 75% of optimum, while Bermuda grass (a warmer species) can only operate at 55% of maximum at the same temperature. Bermuda grass works best at 95° F (35° C). Conversely, many of the cereals, such as northern barley, wheat and rye, photosynthesize best at 50–62° F (10–17° C). Generalizations are hard to come by in the plant world, and there is no single optimum temperature for the universal process of photosynthesis.

When will a plant come into flower? *Saccharum spontaneum* grows wild only between latitudes 5° N and 35° N. Temperature must be an important factor, but temperature as the only factor can be misleading because such a wide range includes many different kinds of climate. With English trees it has been observed that photocontrol has prevented premature bud-opening even though a warm temperature has, so to speak, been encouraging it. The photocontrol prevention is not always successful; contrary weather can confuse many a plant and tree by encouraging spring time growth and then devastatingly, and literally, nipping it in the bud.

The complications implicit in a planetary life can be overwhelming. Think of the confusion that the wide-ranging Saccharum must be in. It is the same species all over this range, and yet temperature, both maximum and minimum, and the likelihood of frost, vary widely in a kingdom spanning 30° of latitude. Daylength also varies, for the southerly point experiences the regularity of tropical daylength and the northern part suffers the relatively short days of winter. What yardsticks is this plant to use as optimum for flowering and how do these change from south to north? How can the same species operate successfully over such a wide range?

Much has been learned in the fifty years since Garner and Allard reported their photoperiodic findings, that *Liliodendron tulipifera* and two other plants continued to grow when the light period was lengthened with artificial light, and ceased upward growth when the days were short. There is far more to be learned, particularly with slower growing woody plants, both about the controlling roles of daylight and daylength, and how these interact with temperature, water and nutrition. As an American botanist said at a recent conference, the regulation process has produced plenty of theories about its nature, and knowledge at present is wide but not deep. Some of that depth should come in the next half century.

Postscript

What will come in the next half century is a better understanding of the adaptations to those cycles of events which have existed for thousands of millions of years. The daily round has always existed, and the years have always been caused by each peregrination around the sun. The canting over of each earthly hemisphere, to face the sun in summer, to face away in winter, has helped to exaggerate the seasonal effects of each yearly cycle. The moon has added its own and independent influence to the two basic revolutions of the days and the years.

Into this hurly-burly of moving parts has come life. It has never been without these gyrations, bringing darkness and lightness, warmth and cold, moisture and drought. To every living thing there is indeed a season. Consequently life has reacted to all this cyclical behaviour of the world about it. Its own seasons have been brought into line with the astronomical seasons. Its own cycles, of mating, of breeding, of rearing, have been made to fit. It also migrates. It hibernates. It aestivates. It changes colour. It has one clock or more. Remember that crab keeping in tune both with the phases of the moon and with the greater cycle of the Earth's journey round the sun.

A world without seasons and without nights and days is unthinkable. To the animals and the plants these cycles are paramount. To all primitive men they were equally crucial. To many today they have lost none of their authority, and even urban man, so artificially maintained, is still ruled by them, sometimes subtly, often blatantly, however much he may resent, deny or welcome the fact. In fact, as planes hustle him across the lines of longitude or the bright lights turn night into day, he could do well to think rather more about the basic cycles of his planet. The various biological clocks are being given greater and greater credit for their roles as pacemakers for life in general. It is strange that man himself is giving less and less thought to his own. To everything there is most positively a season. There are no exceptions, not even for him.

Geological Calendar

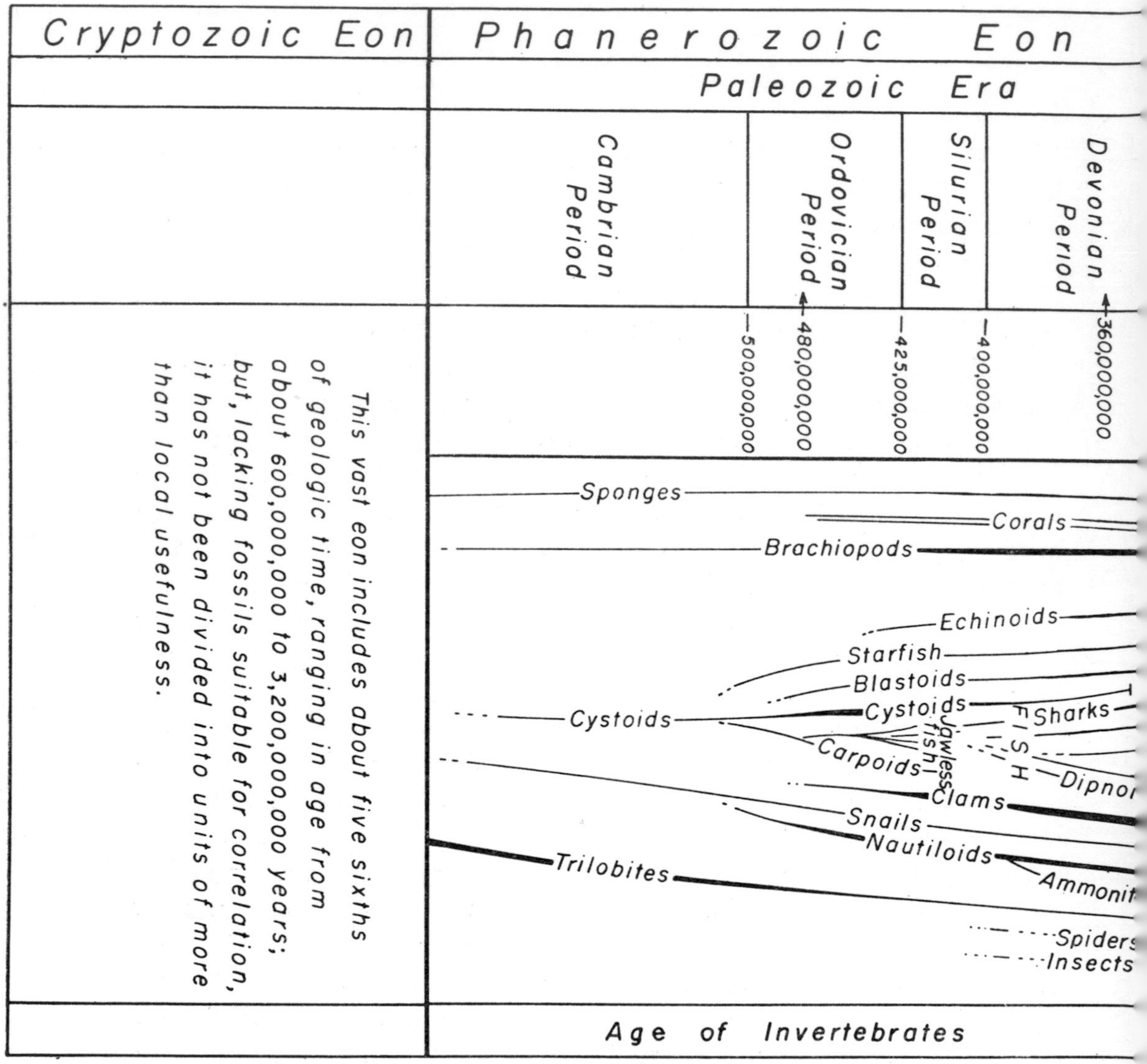

Cryptozoic Eon

Phanerozoic Eon

Paleozoic Era

Cambrian Period

Ordovician Period

Silurian Period

Devonian Period

500,000,000

480,000,000

425,000,000

400,000,000

360,000,000

This vast eon includes about five sixths of geologic time, ranging in age from about 600,000,000 to 3,200,000,000 years; but, lacking fossils suitable for correlation, it has not been divided into units of more than local usefulness.

Sponges

Corals

Brachiopods

Echinoids

Starfish

Blastoids

Cystoids

Cystoids

Sharks

FISH

Jawless fish

Carpoids

Dipnoi

Clams

Snails

Nautiloids

Ammonit

Trilobites

Spiders

Insects

Age of Invertebrates

The geological calendar: the first three columns show the major units of geological time. Column four lists some of the best established absolute dates based on radioactive isotopes; dates closely located in the geological time scale are shown in larger type and the darts indicate their position in column three; dates located approximately on indirect evidence are shown in smaller type. In the life record a dart at the end of a line indicates that a group is still living; a crossbar indicates extinction.

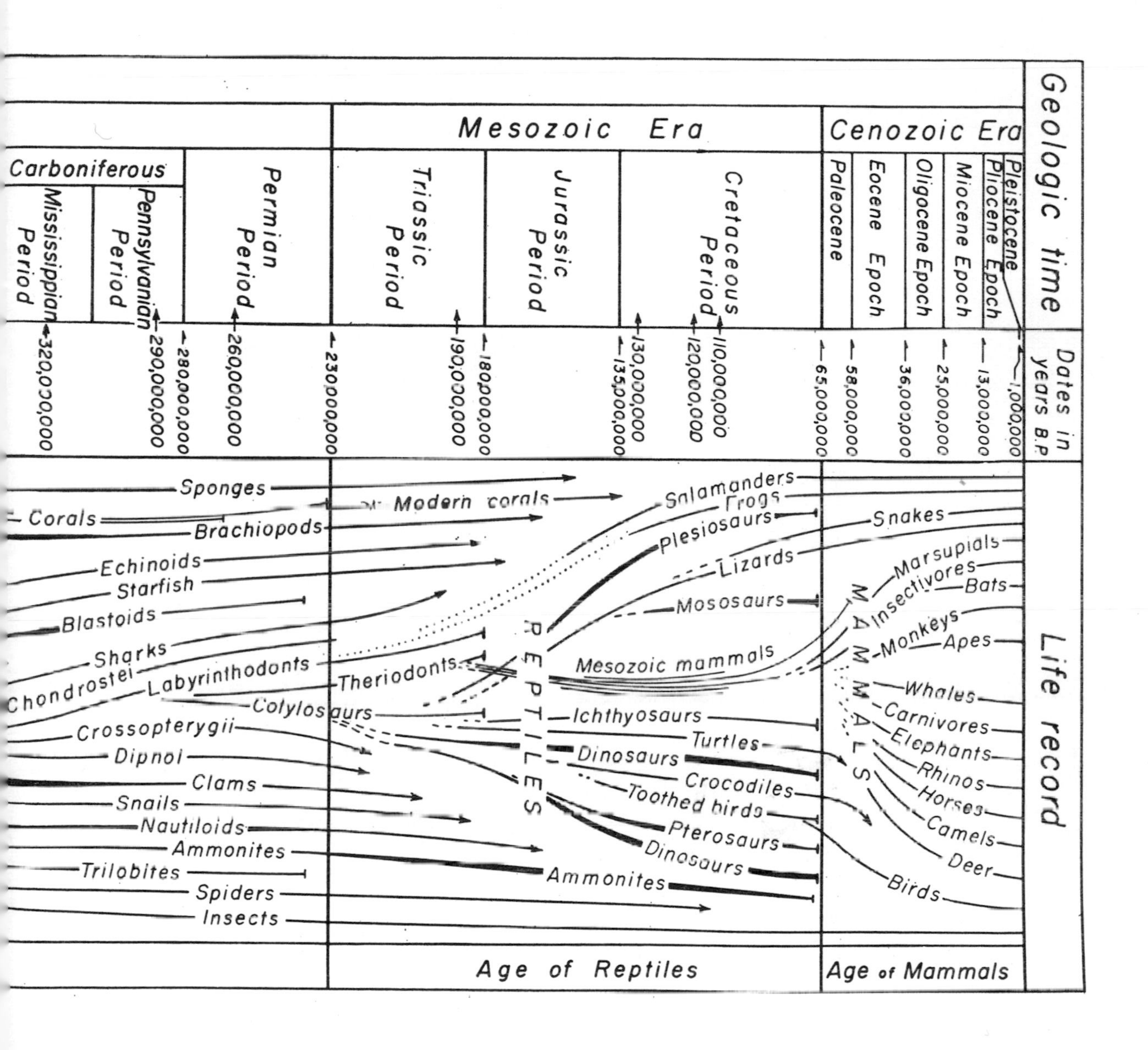

Geologic time
Mesozoic Era
Cenozoic Era
Carboniferous
Mississippian Period
Pennsylvanian Period
Permian Period
Triassic Period
Jurassic Period
Cretaceous Period
Paleocene
Eocene Epoch
Oligocene Epoch
Miocene Epoch
Pliocene Epoch
Pleistocene
Dates in years B.P.
320,000,000
290,000,000
280,000,000
260,000,000
230,000,000
190,000,000
180,000,000
130,000,000
135,000,000
120,000,000
110,000,000
65,000,000
58,000,000
36,000,000
25,000,000
13,000,000
1,000,000
Life record
Sponges
Corals
Modern corals
Brachiopods
Echinoids
Starfish
Blastoids
Sharks
Chondrostei
Labyrinthodonts
Cotylosaurs
Crossopterygii
Dipnoi
Clams
Snails
Nautiloids
Ammonites
Trilobites
Spiders
Insects
Theriodonts
REPTILES
Salamanders
Frogs
Plesiosaurs
Lizards
Mososaurs
Mesozoic mammals
Ichthyosaurs
Turtles
Dinosaurs
Crocodiles
Toothed birds
Pterosaurs
Dinosaurs
Ammonites
Snakes
MAMMALS
Marsupials
Insectivores
Bats
Monkeys
Apes
Whales
Carnivores
Elephants
Rhinos
Horses
Camels
Deer
Birds
Age of Reptiles
Age of Mammals

Index